£7.50

This is the perfect way to start astronomy. This book stirs the imagination, and puts observation in a framework of social activity and personal adventure. It is a technical guide to the sky, full of helpful practical hints. Written in a lively and fresh style it will enthuse, entertain and inform the reader.

- Get to know how to identify the constellations
- Follow the planets on their annual trek among the constellations
- Map the sky and find a comet
- Choose your own telescope
- What to see in a month of lunar observation and a year of stellar observation

# The Sky: A User's Guide

The Pleiades, November 4, 1989, taken with the 18 inch Schmidt at Palomar by Eugene and Carolyn Shoemaker and David Levy. At the intersection of the two arrows is a short trail of a newly discovered asteroid, 1989 VA, found the previous night. This asteroid has one of the shortest orbits of any known object in the solar system, less than two thirds of a year.

# THE SKY:

## A User's Guide

David H. Levy

**CAMBRIDGE**
UNIVERSITY PRESS

Published by the Press Syndicate of the University of Cambridge
The Pitt Building, Trumpington Street, Cambridge CB2 1RP
40 West 20th Street, New York, NY 10011–4211, USA
10 Stamford Road, Oakleigh, Melbourne 3166, Australia

First published 1991
First paperback edition 1993
Reprinted 1994, 1995

Printed in Great Britain by Bell & Bain Ltd., Glasgow

*British Library cataloguing in publication data*
Levy, David H.
   The sky.
   1. Amateur astronomy
   I. Title
   522

*Library of Congress cataloguing in publication data available*

ISBN 0 521 39112 1 hardback
ISBN 0 521 45958 3 paperback

# Contents

| | | |
|---|---|---:|
| | Foreword | xv |
| | Preface | xvii |
| | Acknowledgments | xix |
| **Part 1** | **Getting started** | **1** |
| **1** | **First Nights** | **1** |
| | 1.1   Road map | 2 |
| | 1.1.1   Magnitudes | 4 |
| | 1.2   The Big Dipper key | 4 |
| | 1.3   September suns | 6 |
| | 1.4   Orion | 7 |
| | 1.5   The Milky Way | 8 |
| | 1.6   The planets | 9 |
| | 1.7   Celestial co-ordinates and measurements | 9 |
| | 1.8   The star charts | 10 |
| | 1.9   Starry, starry skies . . . | 27 |
| **2** | **Without a telescope** | **27** |
| | 2.1   Lights | 28 |
| | 2.1.1   Haloes | 28 |
| | 2.1.2   Aurora borealis and australis | 29 |
| | 2.1.3   Zodiacal light and Gegenschein | 31 |
| | 2.1.4   Artificial satellites | 32 |
| | 2.2   The planets | 32 |
| | 2.3   Diversity of the stars | 34 |
| | 2.4   The Sun | 35 |

| | | |
|---|---|---|
| 2.5 | The Moon | 37 |
| 2.6 | Mercury | 37 |
| 2.7 | Planets in daylight | 37 |
| 2.8 | Variable stars | 38 |
| 2.9 | Deep sky objects | 39 |
| 2.10 | Searching | 40 |
| **3** | **Meteors** | **40** |
| 3.1 | Showers | 41 |
| | 3.1.1  Showers month by month | 42 |
| 3.2 | Observing procedure | 47 |
| | 3.2.1  Single observer | 47 |
| | 3.2.2  Group observing | 48 |
| | 3.2.3  Hints | 50 |
| 3.3 | Fireballs | 50 |
| **4** | **Choosing a telescope** | **51** |
| 4.1 | Binoculars | 53 |
| | 4.1.1  Anticipating problems | 54 |
| 4.2 | Telescopes | 55 |
| | 4.2.1  Refractor | 55 |
| | 4.2.2  Reflector | 57 |
| | 4.2.3  Compound telescopes | 57 |
| 4.3 | Eyepieces | 57 |
| 4.4 | Mounts | 58 |
| 4.5 | Why not make your own? | 58 |
| 4.6 | Extremes | 60 |
| **5** | **Recording your observations** | **61** |
| **Part 2** | **Moon, Sun and planets** | **65** |
| **6** | **The Moon** | **65** |
| 6.1 | Why observe the Moon? | 66 |
| 6.2 | The phases | 67 |
| 6.3 | Training project | 67 |
| 6.4 | Day to day notes | 69 |
| **7** | **Moon II: advanced observations** | **76** |
| 7.1 | Crater drawing program | 76 |
| | 7.1.1  Drawing a feature | 76 |
| | 7.1.2  A note about notes | 79 |
| 7.2 | Photographing the Moon | 79 |
| | 7.2.1  At the prime focus | 81 |

7.3    Lunar transient phenomena                              81
    7.3.1    Suspect areas                              83
7.4    Notes on advanced projects                             84
    7.4.1    Lunar height measurements                  84
    7.4.2    Viewing difficult features                 86

8    The Sun                                                   87

8.1    Observing the Sun is dangerous                          88
8.2    Observing projects                                      90
    8.2.1    Daily sunspot count                        90
8.3    Other features on the Sun                               93
    8.3.1    Disk drawings                              94
    8.3.2    Detailed drawings                          96
    8.3.3    Photographs                                96
8.4    Advanced work: hydrogen-alpha filters                  97

9    Jupiter                                                   98

9.1    Jupiter and its moons                                   98
9.2    Seeing                                                  99
9.3    The face of Jupiter                                    100
9.4    Drawing Jupiter                                        102
    9.4.1    Full disk drawings                        102
    9.4.2    Specific regions                          104
9.5    The Galilean satellites                                106

10    Saturn                                                  106

10.1    Historical perspective                                107
10.2    The rings                                             108
10.3    The globe                                             109
10.4    Drawing Saturn                                        109
    10.4.1    A cloudy night experiment for clubs      109
    10.4.2    Actual drawing                           110
    10.4.3    Estimating conspicuousness               110
10.5    Estimating intensity                                  111
10.6    The moons                                             112
    10.6.1    Titan                                    112
    10.6.2    Iapetus                                  113
    10.6.3    Phoebe                                   113

11    Mars                                                    114

11.1    Observing Mars                                        117
11.2    Drawing Mars                                          118
11.3    Kinds of changes to expect                            119

11.4    Surface features                                          120
    11.4.1    The atmosphere                                      122
11.5    Phobos and Deimos                                         123
11.6    Mars thought                                              125

12    Five planets worth watching                                 125

12.1    Venus                                                     125
    12.1.1    Observing Venus                                     126
    12.1.2    Advanced observing                                  127
    12.1.3    Ashen light                                         128
    12.1.4    Transits                                            128
12.2    Mercury                                                   129
    12.2.1    Observing Mercury                                   130
12.3    How the outer planets were discovered                     130
    12.3.1    Discovery I: Uranus                                 130
    12.3.2    Discovery II: Neptune                               131
    12.3.3    Discovery III: Pluto                                133
12.4    Observing Uranus                                          134
12.5    Observing Neptune                                         135
12.6    Observing Pluto                                           136

Part 3    Minor bodies                                            139

13    Asteroids                                                   139

13.1    Historical perspective                                    139
13.2    Naming of asteroids                                       141
13.3    Observing asteroids                                       142
13.4    Kinds of asteroids                                        142
13.5    Observing asteroids                                       142
    13.5.1    A life list of asteroids                            143
13.6    Asteroid occultations                                     145
13.7    Physical observations                                     146
    13.7.1    A photometric study of some asteroids               146

14    Comets                                                      147

14.1    Comets, clouds, and variable stars                        148
14.2    Comet observers                                           149
14.3    What is a comet?                                          150
14.4    Families of comets                                        151
14.5    Groups of comets                                          151
14.6    Observing comets                                          151
    14.6.1    How to estimate the brightness of a comet           153
14.7    The coma                                                  154
14.8    Comet hunting                                             155

14.9    Procedures for hunting                                    157
        14.9.1    Sun vicinity                                     158
        14.9.2    Twilight horizon                                 158
        14.9.3    A group search program                           158
14.10   Hunting with a telescope                                  158
        14.10.1    Search procedures                               159
14.11   Appropriate times                                         160
14.12   Discovery                                                 160
14.13   The naming of comets                                      162

Part 4   Deep sky                                                 165

15    Double stars                                                165

15.1    Mizar                                                     165
15.2    Historical notes                                          166
15.3    Nature of doubles                                         167
15.4    Observing double stars                                    168
        15.4.1    Recording your observations                     169
        15.4.2    Doubles as optical tests                        169
        15.4.3    The Tombaugh–Smith seeing scale                 170
15.5    Advanced work                                             172

16    Variable stars                                              173

16.1    The AAVSO                                                 173
16.2    Eclipsing binaries                                        175
16.3    Cepheids                                                  175
16.4    Long period stars                                         176
16.5    Semiregular stars                                         179
16.6    Cataclysmic variables                                     179
16.7    T Tauri                                                   180
16.8    Naming of variables                                       180
16.9    How to observe a variable star                           182
16.10   Suggested frequency of observation                       182
16.11   Northern summer program                                   183
16.12   Northern winter program                                   185
16.13   A selection of variable stars                            188
16.14   Searching for novae and supernovae                       191
16.15   Neutron star song                                         194

17    The deep sky                                                195

17.1    The *New General Catalogue*                               195
17.2    Open clusters                                             196
17.3    Globular clusters                                         198
17.4    Diffuse nebulae                                           200

17.5   Planetary nebulae                                         203
17.6   Supernova remnants                                        204
17.7   Galaxies                                                  204
17.8   Quasars                                                   207
17.9   Telescope and sky                                         207
17.10  For a city sky                                            207
17.11  For a dark sky                                            210

18     Messier hunting                                           212

       18.1   Messier marathons                                  221

19     The sky on film                                           225

       19.1   Star trails                                        227
       19.2   The Sun                                            228
       19.3   Moon and planets                                   229
              19.3.1   Photography by projection                 229
       19.4   Guided astrophotography                            229
              19.4.1   Camera support                            229
              19.4.2   What you need                             230
              19.4.3   Aligning the polar axis                   231
              19.4.4   Setting up the picture                    233
              19.4.5   Focusing                                  233
              19.4.6   Ready!                                     233
       19.5   Some advanced ideas                                234
              19.5.1   Copying                                   234
              19.5.2   Hypersensitizing                          234
       19.6   Processing film                                    234
       19.7   Some hints                                         237


Part 5   Special events                                          239

20     Solar eclipses                                            239

       20.1   Alignments                                         240
       20.2   Solar eclipses and the public                      241
              20.2.1   Eye protection                            241
       20.3   The saros cycle                                    243
       20.4   Partial eclipses                                   243
       20.5   Total eclipses                                     244
              20.5.1   Photographing a solar eclipse             245
       20.6   Other activities                                   247
       20.7   Annular eclipses                                   248
       20.8   Enjoy it!                                          248

21   Lunar eclipses and occultations                          249

    21.1   Lunar eclipses                                      249
           21.1.1   Shadows                                    250
           21.1.2   Things to do                               251
           21.1.3   Penumbral eclipses                         253
           21.1.4   Thought                                    253
    21.2   Lunar occultations                                  253
           21.2.1   Grazing occultations                       256
           21.2.2   Occultations of planets                    256
           21.2.3   Occultations by planets                    256
           21.2.4   Murphy's Law and occultations              257

Part 6   A miscellany                                          261

22   Passing the torch                                         261

           22.0.1   Schools                                    261
    22.1   Methods of teaching                                 262
    22.2   The planets                                         263
    22.3   Daytime observing                                   263
           22.3.1   Observing the Sun                          264
           22.3.2   Venus                                      264
           22.3.3   Observing the Moon                         265
    22.4   Night observing                                     265
    22.5   Meteors, and learning through research              266
    22.6   Closing thought                                     268

23   The poet's sky                                            268
24   Resources                                                 273

    24.1   Societies                                           273
    24.2   Literature                                          276
           24.2.1   Observing assistance                       276
           24.2.2   Annual guides                              278
           24.2.3   Star atlases                               278
           24.2.4   Historical                                 278
           24.2.5   Solar system                               279
           24.2.6   Deep sky                                   280
           24.2.7   General assistance                         280
           24.2.8   For children                               281
           24.2.9   Textbooks                                  281
           24.2.10  Magazines                                  281

       Index                                                   283

For my brothers Richard and Gerry, and for my sister Joyce

# Foreword

Astronomy is a science for everyone. Anyone who looks at the sky and contemplates its wonders is an astronomer. Observing the sky can be done in as simple or complex a manner as one wants. Armchair astronomy is practiced by many using paper, photographic, and electronic media. Casual practitioners of the observing arts observe the sky first-hand with their eyes, perhaps augmented with binoculars or small telescopes and maybe a camera. Serious observers combine the hardware and software of the other groups and become experts in their chosen fields.

Observers can also be categorized by how they treat their efforts: the diversity of the sky brings out the scientist in some, the accountant in others, and the poet in those who are especially stirred by its beauties and mysteries. This may seem an odd way to describe the efforts of amateur astronomers as they make their observations. Indeed, scholarly papers have been written arguing for classification based on whether one does recreational astronomy or astronomical research as a hobby or as a professional. These and other classification schemes are not mutually exclusive and, in fact, even the classes within them are not exclusive: most observers fit into more than one category.

Consider the amateur scientist doing astronomy. While the 'observing' this person does may be on a computer, it more likely involves keeping detailed records of objects studied at the telescope. The telescope and eyepieces used, weather conditions, general comments, and perhaps a sketch are recorded in a logbook. Those observers with a passion for adding to science and an enjoyment of the suspense and excitement of astronomical mysteries make observations as prescribed by the numerous amateur organizations collecting data for professional astronomers. Their observations are submitted with a deep sense of satisfaction for the role they play in solving the mysteries of the universe.

Other observers keep a tally of the objects they have seen. Some may work through some of the numerous catalogs – specialized or general – of the objects visible in the sky. Others keep a list of the comets they have seen, the variable star observations they have made, the number of telescopic meteors seen, or even the number of observing sessions they have had. As they observe, these skywatchers hone their skills and see more each time they look through their telescopes.

The poetic observer has perhaps the easiest time doing astronomy. He or she can simply stare up at the celestial dome and enjoy the wonder of the scene with a notebook or sketchbook in hand, otherwise unencumbered by the paraphernalia often needed by other observers. This is not to say that poets can't use telescopes, but rather to say that they enjoy more freedom to choose how to observe than the others.

This book offers something for all types of astronomers. The love of the sky conveyed here by the poet combined with the expert's knowledge gained through years of experience in numerous aspects of this rich science provide lessons for everyone interested in the sky.

As you read this book, either straight through or dipping in and out as the mood strikes you, let your imagination and interest run free to enjoy, to learn, and then to experience more of what astronomical observing has to offer. Test yourself and your instrument by pushing the limits of their capabilities. Experiment with new observing techniques to observe more of the universe. Don't limit your astronomical experience to one area of the subject: sample the smorgasbord of wonders the cosmos has to offer. The universe offers an infinite variety of facts and mysteries, and we have only a little time to know and understand and enjoy them. There's no time to lose!

Steve Edberg

# Preface

This is your owner's manual for the sky. Do you know how, when you buy a new product, the instructions often begin by congratulating you on purchasing it? In a sense, you have purchased your interest in astronomy as an article of faith, hoping that by joining a club, subscribing to a magazine, or buying a book like this one, your hours under the stars will be beneficial. So, congratulations! You have above you the only brand of sky in town, and I hope you enjoy it.

Designed to give you a sense of what observing *as an amateur* is all about, this book provides a set of observing suggestions for a number of the many fields of observing. Every area offers complexity and challenge, but in different forms. There are at least two particular ways to observe 'stars', for example. Variable star observing offers greater scientific reward for an amateur than does double star observing, but each field takes time and patience to develop, and I think that both are equally fun.

If you are interested in observing only to help the professionals, you may be in amateur astronomy for the wrong reason. The first rule of observing is that it be satisfying. For the most part, professionals are not greatly interested in our work, although good observations in any area might attract their attention.

Observational astronomy offers something for everyone. Each observing discipline has its own code of requirements, both scientific and social, and you should be able to find one that closely matches your interests and temperament. For example, meteor shower observing is often done in groups of six or more to ensure adequate coverage of the entire sky. This encourages meteor shower 'parties' that can last all night long, leaving the participants intoxicated, as it were, with meteoric dust. Eclipses also are fine gatherers of astronomy enthusiasts, as observers like to share the awesome moments of these celestial line-ups.

Where meteor observing ought to be the grand excuse for an ultimate celestial party, comet hunting is a loner's pursuit. Rather competitive, it is almost always done by an observer working alone. The person who leaves a snug bed for a 3 a.m. look at Mars, because that is when Mars is highest in the sky and thus most clearly seen, enjoys a special communion with a planet that is not quite the same when a crowd of sightseeing stargazers congregate nearby.

This book is designed to suggest enjoyable areas of observation. It is slightly weighted toward solar system observation, since objects in this region are more easily seen from the urban locations most beginners observe from. Not a textbook, its facts relate directly to your observing needs. Here you will pick up ideas on how and what to observe; this book guides you into more productive and happy use of your observing time. But these are just ideas; later you might come up with original projects to make your observing time even more interesting. Whether for socializing or for science, observing is an activity that can rapidly become your outlet to relax and to become involved in a fascinating avocation.

Each part of this book has been organized according to how easy or difficult objects are to observe. The planets, for example, begin with Jupiter and progress through the objects which, through brightness or times of visibility, are not as easy to watch on a regular basis.

When you buy something, you expect a guarantee. There is none for the sky. Some once-active observers I know have lost their enthusiasm for observing, and I must admit that in the 8200 times I have been under the stars, there were two or three times when I was not overwhelmed with what I saw. You may lose interest; many do. I do hope that your interest lasts, that it is strong and satisfying, and that you do your best to spread it to someone else.

Amateur astronomy is not just a hobby, it is a commitment, a way of life that transcends the detail of daily existence. Look through your telescope thoughtfully tonight, for it is more than starlight that its mirror will reflect. From the unknown dimensions of space and time will return also a part of yourself.

# Acknowledgments

To Stephen J. Edberg, for his thorough review of every word of the manuscript, and to his son, Aaron, I am in great debt. Steve took the time to read two or more drafts of several chapters.

Walter Haas, founder of the Association of Lunar and Planetary Observers, offered general advice. John Westfall, Director of the ALPO, reviewed the two Moon chapters and the material on eclipses. Gary Rosenbaum examined the Mars chapter; Rik Hill evaluated the Sun chapter; and Dolores Hill made many suggestions on the meteors chapter. Charles Morris and Jean Mueller were most helpful.

Larry Stein, Richard Levy and Peter Jedicke read through the introductory chapters. Derald Nye evaluated the material on occultations; Paul Lorenz gave advice on the deep sky chapter; Judy Stowell typed early versions of the manuscript; and Lonny Baker thoroughly reviewed the solar eclipse chapter. Peter Jedicke prepared the index.

Parts of Chapter 23 were adapted from the first two chapters of my 1979 Queen's University M.A. Thesis, 'The Starlight Night: Hopkins and Astronomy'; an article 'Poet and observer: Gerard Manley Hopkins and some mid-19th century comets' in *The Journal of the Royal Astronomical Society of Canada*, vol. 75, no. 3, 1981; and my 'Observer's Cage' column in the October, 1986, National Newsletter of the RASC. I thank Norman MacKenzie of Queen's University for his inspiration and help with the material that appears in this chapter. Simon Mitton and Cambridge University Press were very friendly and helpful through every stage of the process of the book's publication.

I am thoroughly indebted to all these people for their time and effort to help make this a thorough and accurate observing book.

# PART ONE

# Getting started

## 1  First nights

One cold spring night back in the middle 1950s, my older brother rushed in from outside. 'I'm sure that's it!' – his eyes were wide with excitement. 'The Big Dipper!' I had no idea what a big dipper was, let alone any notion of what role such a thing might play in my later life, but if someone had really gone to the trouble of placing one in the sky, I wanted to see it. I rushed to the front door and Richard pointed out the seven twinkling lights.

Together we watched the stars for a few minutes until one of Richard's friends dropped by. 'Look up,' Richard said, 'the Dipper!' The friend looked briefly to the north and then back at us. 'So what?' she answered. 'I've seen it thousands of times.'

That was my first look at the stars. And it could very well have been the last, except that the image of the Dipper stuck in my mind until the following summer when I was trying to fall asleep during my first lonely time away from home at summer camp. Once again I saw the stars, but this time there was no one else to show them to me, no dippers or other standard shapes to learn. For the first time I felt free with the stars. Yes, they seemed far, very far, away, but I could make them into any shape I chose. Maybe I wasn't so far from home after all.

My earliest looks at the stars typify the two most popular ways of meeting the sky. The organized star party, during which a trained observer points out all the constellations from Auriga to Vulpecula, is traditional and accepted. But unless it is handled with care and wisdom, the long lines and all too brief looks can ruin your interest in the stars. The second way allows you to meet the stars on your own terms. My lonesome night was relieved not by mytho-logical figures but by real stars. That night was not a star party but a communion.

You don't have to wait for someone else to show you the stars. Go out one night and look patiently at those distant lights, and ask questions. Are they all white, or do some show hints of color, bluish or reddish hues that might tell us something about their nature? Are some parts of the sky more crowded than others? Do certain bright 'stars' shine with a steadier light than the shimmer of the rest? And finally, what constellations, what shapes, can *you* invent from the stars?

Your first night out should be a voyage on a magic carpet that takes you to other places and other times. Even a casual look at the stars gives you a share in the company of the timelessness that they represent. As you become more experienced and feel at home with the stars, you will probably want to get to know them better through the eye of a telescope. As you read through this book, your view might become more oriented toward science, but please, never let the romance of the sky disappear from your mind.

Although the stars are as common as flowers and trees, most people cannot name a single star other than the Sun. Most likely this is because the more earthly varieties of nature are visible during the day, while stars come out at night when people are indoors.

You will begin your skywatching from one of two kinds of locations: a city-bound site, perhaps with limited horizons, from which less than a hundred stars are visible, or from a site out in the country where the sky is splattered with possibly 3000 stars.

Actually, a *city* site is better for starting out, because of its limited number of stars. Those who are blessed with a dark site won't appreciate it until later, when a little knowledge will ease their way around. At first, dark site users actually have a harder time. Their sky is so filled with stars that picking out a few bright ones is confusing; they need to wait for a moonlit night to start!

## 1.1    Road map

In this chapter I treat the sky as a series of roads and streets, needing a map to navigate it. The charts included are intended just to get you started. Afterwards, you should invest in a more detailed atlas of the sky. The bibliography (Chapter 24) suggests some atlases which may help at different levels, as well as magazines which bring you up to date on where the bright planets are, and what special events are happening.

Before beginning, it is good to have an idea of what you're going to see out there. That depends on the following factors:

**Latitude** Depending on how far north or south you are from the equator, different groupings of stars will be presented in different ways. For example, from north latitude 45 degrees, the Big Dipper is always in the sky, day and night, all year long. We say that the Dipper is *circumpolar* at that latitude. If you look toward the north, and if you have a good horizon to the northeast and northwest, you should have no trouble finding the Dipper. In autumn and winter it will be low in the sky during the evening hours; in spring and summer it will be very high, though always in the northern sky.

From lower latitudes the Dipper is not circumpolar. At 30 degrees north latitude or less, you cannot see it in autumn and winter, although it is prominent at other times. At 45 degrees south latitude the Dipper is not visible at all; it never rises above the horizon!

If latitude definitely affects what you see, *longitude* does not. Evening in London, England, will show you the same sky as would evening in Calgary, Alberta, but about seven hours earlier.

**Time of night** Because of the Earth's rotation, everything appears to revolve around the north and south celestial poles. Fortunately, we can find the north celestial pole because a bright star called Polaris lies within a degree of it. The southern hemisphere does not have a bright pole star, although a line joining Gamma and Alpha Crucis and extended about 4.5 times that distance will pass close to the south celestial pole. The nearest equivalent to a polar star there is a fifth magnitude star, barely visible to the naked eye, called Sigma Octantis.

Incidentally, these stars are designated by a system of 'Bayer letters' devised by Johann Bayer for his *Uranometria* atlas of 1603. Generally these letters are assigned in order of brightness, so that Alpha Crucis is the brightest star in the Cross. As we proceed, we will find many examples of stars referred to by their names (Vega, Rigel) as well as their Greek letter designations (Alpha Lyrae, Beta Orionis). The Latin genitive form of the constellation is used in these cases, so that Gamma of Leo is called Gamma Leonis.

Even as the Sun rises and sets, it actually appears to circle the poles in an arc that changes depending on the Earth's place in its orbit. The other stars all appear to move around the poles in one day, so that the stars we see in the east in the evening are high in the sky at midnight, and low in the west near dawn.

**Time of year** The day the stars follow, however, is not the 'solar day' of 24 hours that we know. Because the Earth also moves a bit in its orbit each day, the stars appear to move across the sky in a 'sidereal day', which is about 3 minutes, 56 seconds faster than a solar day. Although this doesn't seem like much, it adds up quickly, almost to half an hour a week, and to two hours each month. Thus, a star that crosses the meridian, the highest point in its path across the sky, at 9 p.m. tonight, will cross it tomorrow night at 8:57:04, and next month it will cross around 7 p.m.

**Phase of Moon** Even in the smog and light-polluted sky of a city, there is a difference between what one can see at new Moon compared to full Moon. The light of the full Moon is very brilliant, swamping out the light of all but the brighter stars. It is definitely something to take into account when planning an observing session from a dark site. On bright moonlit nights you should not plan to observe the faint nebulae, galaxies and comets; these objects require a dark sky background to be seen best. (Incidentally, the Moon loses half its brightness only two days after its full phase!)

**Weather conditions** A thin layer of cirrus cloud has a very definite negative effect on what you can see. Also, a strong breeze makes everything uncomfortable, from blowing the telescope around on its mount to disturbing star charts. Remember warm clothing for cold weather, and insect repellent for warm summer nights. Remember also that, in many climates, a warm day might be followed by a cold night, so always take along extra clothes.

### 1.1.1   Magnitudes

Knowing that all stars and solar system objects are not the same brightness is one of the first things anyone can understand about the sky. The Sun is so bright that we cannot gaze at it unprotected with the naked eye, without suffering eye damage. The Moon at full phase is bright enough to affect the visibility of everything else in the sky. Venus is so bright it can cast shadows. Several stars are very bright, many more are fainter, and a huge number are barely visible.

We differentiate among star brightnesses by assigning them an arbitrary magnitude. The Sun shines at magnitude $-26.5$, while the full Moon is about $-12$. At greatest brilliance, Venus is $-4$. The bright star Vega is very close to 0 magnitude, and Deneb is 1.6. The faint stars barely visible to the average naked eye under a dark sky might be sixth magnitude.

Each magnitude number is 2.5 times fainter than the next lower one. Thus, a little multiplication will show that a sixth magnitude star is exactly 100 times fainter than a first magnitude star.

You should note that while these numbers are very good guides to how easily visible a star or planet is, they are not as accurate when describing the visibility of extended objects like comets and galaxies. Consider two galaxies, each listed at seventh magnitude; one spreads out over 1 degree of sky (equivalent to two full Moon diameters), the other over $\frac{1}{4}$ degree. Since the first galaxy's light is spread out over a larger area, it will appear much fainter than the other one; the first galaxy has a lower 'surface brightness' than the second. If you read that a new comet will become fifth magnitude, don't expect to see it under a suburban sky without a telescope or powerful binoculars. Because comets often have low surface brightnesses, they usually need to be third magnitude before they become easy to see from a city location.

## 1.2   The Big Dipper key

For northern hemisphere observing, the Big Dipper can get us started on constellations all over the sky. Here are some examples:

**Polaris** The two stars at the end of the bowl opposite the handle can be joined with an imaginary line. If you extend that line about five times their distance, you find Polaris, the North Star. This star is at the end of the tail of *Ursa Minor*, the Little Dipper, a faint group of stars that is not easy to find in a city sky.

**Bootes, Corona Borealis, and Virgo** The Dipper's handle is a curved line. From February through to August, by extending the arc, you can 'arc to Arcturus', the brightest star in the constellation of *Bootes*, the Herdsman, a formation that resembles a kite. Just to the east of this kite is a semicircular formation called *Corona Borealis*, the Northern Crown, consisting of a fairly bright star (Alphecca) and several fainter ones.

   The arc to Arcturus trick continues to work if you simply continue the line past Arcturus and 'speed to Spica', the brightest star in the constellation of *Virgo*. Although Arcturus and Spica are approximately the same brightness, they have very different temperatures, a fact you can verify any time you see them from your back yard. Arcturus is reddish, indicating a relatively cool star, and Spica is bluish, signifying a hot one.

**The summer triangle** Join the two inner stars of the bowl with a line that you continue northward. It will eventually reach a large area of sky punctuated by three bright stars: Vega in *Lyra*, the Harp; Deneb in *Cygnus*, the Swan; and Altair in *Aquila*, the Eagle. The Harp and Swan look like what they are supposed to look like; the four faint stars near Vega form a parallelogram that you can almost believe is joined by harpstrings. Cygnus resembles more a cross than a swan, with Deneb at the top. The swan image works if you see Deneb at the tail (the meaning of the name Deneb) and the long neck pointing between Vega and Altair to Albireo, one of the sky's most famous double stars (see Chapter 15). The arm of the cross, extended on either side, becomes the wings of a graceful swan in flight. Aquila the Eagle consists of a grouping of stars out of which, with a little imagination, you can see an eagle.

   The summer triangle dominates the sky from June to December. For children, I like to add that it also forms the letter V, which stands for vacation; when they see the V rising in the east on June evenings, they know that summer 'V'acation is not far behind. Southern hemisphere observers will generally not see much of this triangle, although it could presage winter vacation for those students!

**Scorpius and Sagittarius** A line drawn from Gamma Ursae Majoris, the star at the bottom of the bowl on the handle side, through Eta, the end star in the Dipper's handle, will point across the sky toward Antares, the bright red star in *Scorpius*, a constellation that actually looks like a Scorpion. To the east of Scorpius is *Sagittarius*, the Archer. These two southern constellations are best visible from the northern hemisphere on evenings between May and August.

From December through to March, the Dipper's same two stars point in the opposite direction to one more prominent constellation. A line from Eta through Gamma leads across an expanse of sky now to Procyon, the brightest star in a small constellation called *Canis Minor*, the Little Dog.

From September through to March, try a line beginning at Eta, the end star of the handle, and continuing through Zeta, the handle's middle star. (This star is also known as Mizar, and is a famous double star.) Extend that line across the pole to Capella, the bright yellowish sun in *Auriga*, the Charioteer. Can you see three faint stars close to the southwest of Capella? With this triangular asterism, Capella is known as the Goat and the Kids, a group whose evening rising in November was seen in ancient times as a harbinger of the winter storm season.

Between January and June, the Dipper can be used in a somewhat unconventional way to point towards *Leo*, the Lion. Fill the Dipper's bowl with water and cut some small holes in the bottom. As the water sprinkles out, you can find Leo taking a shower! (If you live in the southern hemisphere, you might have a difficult time imagining Leo showering in this way. If the Dipper appears at all, the Lion appears above its upside down bowl! A line from Alpha through Gamma Crucis and extended far northwards will meet Leo.)

## 1.3    September suns

Since the Dipper is not too helpful at this time of year, its position in the direction of the Sun making it difficult to see, you are more on your own. However, you do have the Great Square of Pegasus, an asterism that is favorable to both hemispheres.

The Square consists of four moderately bright stars, each about 2.5 magnitude. The one at the Square's northwest corner forms a triangle with two other stars nearby. Since the Square is a little north of the celestial equator, it is seen about midway between the north and south poles. One of the Square's stars is actually not part of Pegasus: Alpha Andromedae (or Alpheratz) is the northeast corner star. Finding other constellations from the Square is fun, especially if you keep a version of their mythological story in mind: Andromeda, the daughter of Cepheus and Cassiopeia of Ethiopia, was rescued from Cetus, the Whale, by Perseus, who carried Andromeda into the sky on Pegasus, the Winged Horse.

Andromeda is a constellation of faint stars to the northeast of Pegasus. It is found by joining Alpha Pegasi (southwest corner) and Alpha Andromedae (northeast); the line extends into Andromeda. Extending the Alpha–Alpha line the same distance will lead to Perseus. A line joining Gamma Pegasi (southeast corner) and Alpha Andromedae (northeast) and extended about four times that distance will lead to Cassiopeia, the Queen, a bright constellation shaped roughly like the letter W or M. (Alternatively, the line from the pointers in Ursa Major to Polaris, if continued about the same distance, will also lead to Cassiopeia.) To find Cepheus, the King, join the two bright stars at the west end of Cassiopeia, Alpha and Beta, and continue that line about three times the distance to wind up in the middle of Cepheus, a faint constellation that resembles a child's drawing of a house.

Cetus the Whale comprises stars so faint that the constellation is difficult to recognize. A line from Alpha Andromedae (northeast corner) through Gamma Pegasi (southeast) and continued the same distance will go through the 'Circlet' of faint stars belonging to Pisces, the Fish, and the same distance again will end in the middle of Cetus the Whale.

## 1.4　Orion

The key to the December through March sky is definitely Orion the Hunter, whose three second magnitude belt stars shine in a row just south of the celestial equator. Joining them and extending the imaginary line north-westward about five times its distance will land you just south of the bright red star called Aldebaran. This star marks the end of one arm of a V-shaped grouping, actually a star cluster called the Hyades. Can you see the Pleiades cluster to the north?

Now try extending the belt line almost three times its distance to the southeast. The very bright star you will almost meet is called Sirius, and the stars surrounding it are part of Canis Major, the Greater Dog. Next, join Orion's two brightest stars, Rigel (Beta Orionis) on the constellation's southwest side, and Betelgeuse (Alpha Orionis) at the northeast corner, and go northeastward until you reach two almost equally bright stars. The north one is Castor, the south one Pollux, and the constellation they lead is Gemini the Twins. Joining Kappa Orionis at the Hunter's southeast corner with Betelgeuse, and going north about twice the distance, will lead to Auriga the Charioteer, whose brightest star is Capella.

Other patterns are possible. Remember the northern hemisphere's summer triangle? In winter we have a huge 'heavenly G' that begins with Aldebaran and circles through Capella, Castor, Pollux, Procyon, Sirius, and Rigel, before finally turning 'inward' to complete the G with Betelgeuse.

In the southern hemisphere, we can use Orion to point toward other constellations. A line from Betelgeuse through Kappa Orionis, extended about twice that distance, will point to Canopus, the brightest star in Carina. The constellation lies almost entirely to the east of its brightest star.

**The Southern Cross** One of the brightest constellations in the southern sky is Crux, the Cross. I have already mentioned its use in pointing toward the south pole. By continuing the Crux-to-pole line, you should meet Achernar, the brightest star and mouth of the long and faint constellation of Eridanus, the River. Join Delta Crucis (west star) and Beta Crucis (east) to point towards two very bright stars, Beta and Alpha Centauri. At 4.3 light years away, Alpha Centauri is the closest star to the Sun. Join Beta Crucis (east star) and Alpha Crucis (south) to meet the beautiful constellation of Carina, with bright Canopus on the constellation's far west side.

These pointing suggestions should be enough to get you started. As you become more familiar with the stars, you will invent your own aids to lead you to more new regions.

## 1.5    The Milky Way

One of the most beautiful things in the sky, and one of the most confusing, is the Milky Way. Invisible from most city locations, and bright as a cirrus cloud under a dark sky, the Milky Way arches through the sky like a subtle feather.

Actually, the Milky Way does not look like a cloud at all when seen through binoculars, which reveal it for what it is, a swarm of faint stars. Here is where most of the stars in our galaxy lie; when we look at the Milky Way, we look along the plane of our Milky Way galaxy.

We believe our galaxy to be shaped like a giant pinwheel: flat, with arms that spiral out from the center. If we could look at this galaxy from a great distance we would have no difficulty discerning this shape; you can do this any night by looking at the galaxy in Andromeda known as Messier 31. In our own case, we are looking from within, as if we were trying to visualize the shape of a house with cloudy walls when all we can see is one or two of the inside rooms, and only with a dim view of the rest.

It is our galaxy's flattened shape that causes the Milky Way to appear as a single band of light instead of a general, sky-covering swarm. Additionally, we are off in the boondocks, in one of the spiral arms. If you want to look toward the galaxy's center, look toward Scorpius and Sagittarius. On a dark night you will notice that the Milky Way widens and brightens there; that is where the center is.

The Milky Way is not a smooth band of light, you will quickly notice. It has different consistencies and 'holes', and in the middle of Aquila it splits into two, in a division so apparent we call it the 'Great Rift'. The cause of the rift and the holes is our galaxy's dark matter, enormous amounts of gas and dust whose presence we only infer by seeing the numbers of stars decrease behind it.

Although this strange shape of the Milky Way has been obvious ever since people started looking at it, it was only with John Herschel's nineteenth century observations that the extent of these 'holes in the sky' became a subject of interest. Besides the Great Rift in the northern Milky Way, perhaps the most stunning example of dark matter is the Coal Sack Nebula, not far from the Southern Cross.

## 1.6    The planets

Just as you are sure you have the brighter stars memorized, a bright interloper will come along and confuse your carefully perceived star patterns. Chances are this interloper is a planet, so named because it is not a 'fixed star' but a wandering object. If you have a telescope – any telescope – set it up and see what it does to the planet. It may show the object in a phase like the Moon, it may show nearby objects that might be moons, or it may show rings. My first look at a planet through a small telescope happened just this way; I did not know what planet it would be, and I will never forget my amazement when my telescope revealed the rings of Saturn.

Planets will have high priority on your observing list! Among other activities suggested in the next chapter, following the planets on their annual treks among the constellations can be an engrossing activity.

## 1.7    Celestial co-ordinates and measurements

The comet hunter Leslie C. Peltier once wrote of the excitement of going from figures on a printed page to a comet in the sky. To share that excitement, we need co-ordinates to read the sky's road map. Like the Earth, the sky has been outfitted with a set of co-ordinates called *right ascension*, corresponding to longitude, and *declination*, corresponding to latitude. Although you do not really need to know much about these to get started in astronomical observing, an understanding about how they work will aid your searching for new objects, especially when you use star atlases.

Right ascension is based on a star's apparent sidereal movement from one night to the next. Thus, the sky has been artificially divided into 24 hours of right ascension, and each hour successively into minutes and seconds. This one co-ordinate of an object might be listed as RA 18h 47m 22.7s.

The sky has also been divided into degrees of declination with zero at the celestial equator and continuing northward to +90 and southward to −90. Each degree is divided into 60 minutes, and each minute into 60 seconds, so that an object's declination could be 13 degrees 14m 43s. At the equator, a degree of declination is exactly the same as four minutes of right ascension; you can test that by looking at a star on the celestial equator through your telescope, with an eyepiece that gives a one-degree field. It will drift through your eyepiece field of view in four minutes. (However, if the star is not near the celestial equator, the four minutes it drifts will not correspond to what you see in your eyepiece field of view.) Remember that minutes and seconds of right ascension, known as minutes and seconds of time, take up about 15 times more sky than do those of declination (minutes and seconds of arc).

When you see these co-ordinates printed in a sky atlas, you will also see a year attached to them, usually 2000 or 1950. This is the *equinox* to which the

charts are set. It is not something you need to worry about unless one of two
things happen: one, you get a telescope mount with setting circles; and, two,
a comet is found and you try to locate it in your telescope using the printed
positions given in a magazine.

Equinox 2000.0 means that the charts are exactly right from 1999.95 to
2000.05. They change slightly because of a slow top-like wobble of the
Earth caused by the Moon's gravitational pull. It is the same effect that
causes the apparent position of the celestial poles to change in a period of
26 500 years; when the northward-pointing pyramids were aligned toward
the pole, Thuban in Draco was there, and years from now Vega will be our
Pole star.

If you use your telescope's setting circles, they will be accurate to equinox
1990.2, or whatever the date you are observing is. For atlases set to 2000.0,
the correction is very small and not really to be worried about. If your atlas is
set to another of the popular equinoxes, like 1950, 1900, or 1855, the
correction is far larger.

When a newly found comet or nova is announced, its positions are
published to a particular equinox, like 1950.0 or 2000.0. Say that on February 25, for example, you observe a new comet at RA (2000.0) 19h 40.8m,
Dec (2000.0) $-15.6$ degrees. If your atlas is also set to equinox 2000.0, you
have no adjustment to make; simply plot the position of the comet on the
atlas using the co-ordinates, and look for it. If the atlas is set to 1950, the
positions will be off, and conversions take some mathematics or a computer
program. It is far easier to have everything in the same equinox.

Your hands form an easy way of measuring distances in the sky. At arm's
length, the distance across your clenched fist is about 10 degrees; from
thumb to fifth finger of fully separated fingers is about 25 degrees. With or
without a knowledge of co-ordinate systems, you are now ready to embark
on a more serious celestial journey. The next chapter will suggest some
activities that require neither binoculars nor telescope.

## 1.8    The star charts

The following charts are designed to show the sky in two different
ways; first, from a suburban location from which not too many stars are
visible; and, second, from a darker sky showing many more stars. A beginner
should use the suburban set first; otherwise the sheer number of stars might
cause confusion. Although observers at latitude plus 40 degrees or minus 30
degrees will find the charts most effective, a few degrees north or south will
affect only the visibility of stars near the horizon.

The ideal observing time is 9 p.m. local time. Again, an hour earlier or
later will affect only the visibility of stars near the horizon.

The following charts (Figures 1.1–1.16) were prepared especially for this
book by cartographer Robert Miller.

Figure 1.1. Northern winter evening sky as seen from a suburban location. Stars shown to magnitude 4.3.

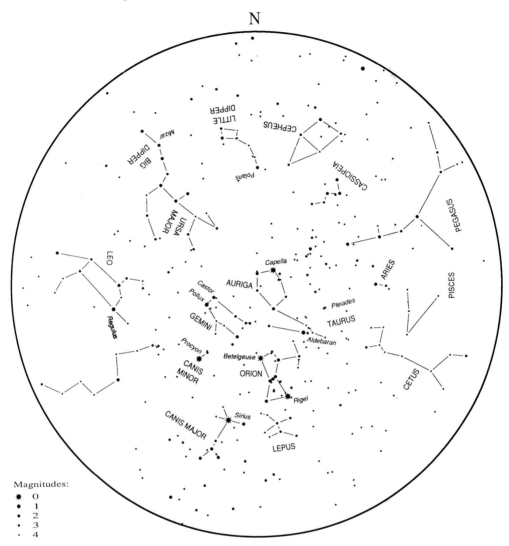

Figure 1.2. Northern spring evening sky as seen from a suburban location. Stars shown to magnitude 4.3.

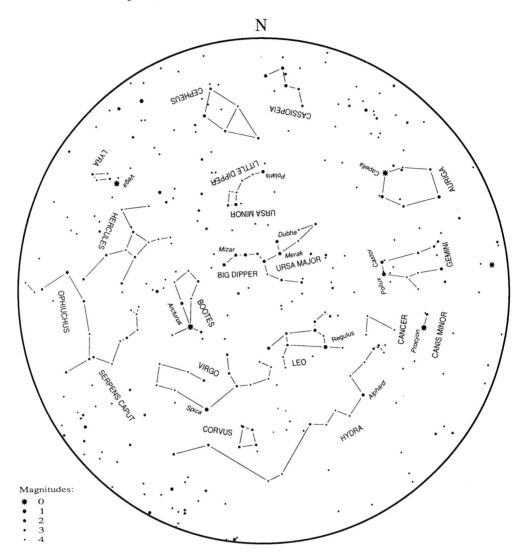

Magnitudes:

✹  0
●  1
•  2
·  3
·  4

Figure 1.3. Northern summer evening sky as seen from a suburban location. Stars shown to magnitude 4.3.

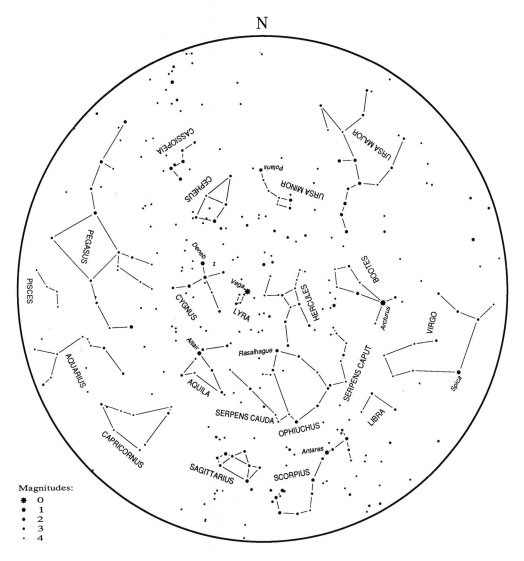

Magnitudes:
✳ 0
● 1
• 2
· 3
· 4

Figure 1.4. Northern autumn evening sky as seen from a suburban location. Stars shown to magnitude 4.3.

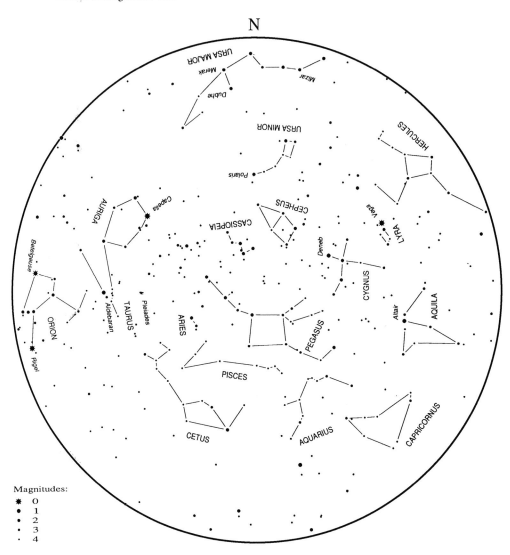

N

Magnitudes:
✶  0
●  1
●  2
·  3
·  4

Figure 1.5. Northern winter evening sky as seen from a relatively dark location. Stars shown to magnitude 5.0.

N

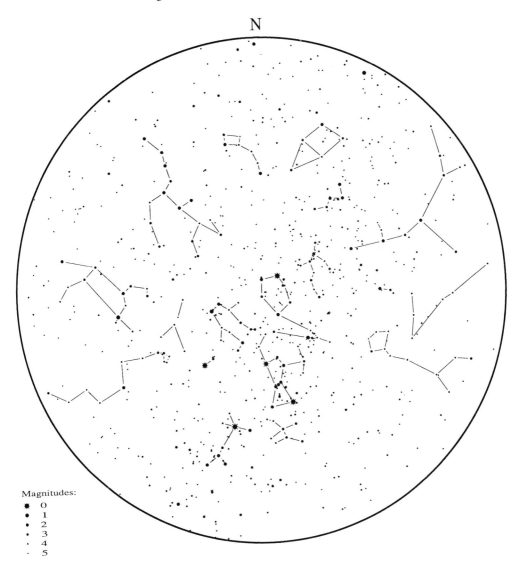

Magnitudes:

* ✸  0
* ●  1
* ●  2
* ·  3
* ·  4
* ·  5

Figure 1.6. Northern spring evening sky as seen from a relatively dark location. Stars shown to magnitude 5.0.

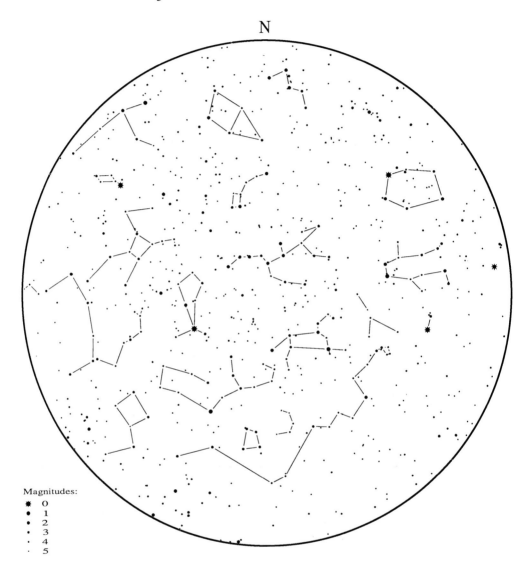

N

Magnitudes:
* 0
● 1
● 2
• 3
· 4
· 5

Figure 1.7. Northern summer evening sky as seen from a relatively dark location. Stars shown to magnitude 5.0.

N

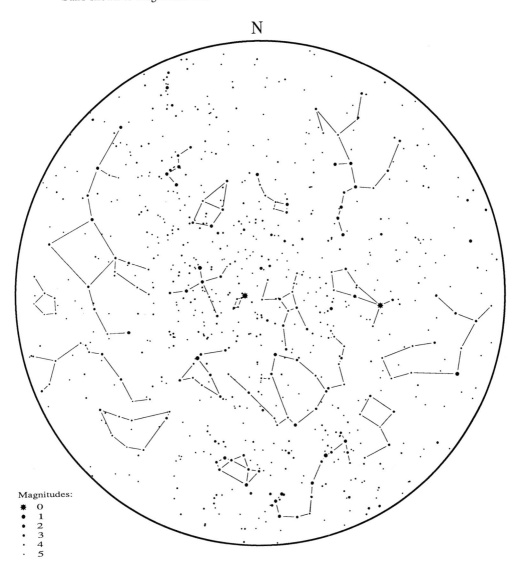

Magnitudes:
✳ 0
● 1
● 2
• 3
· 4
· 5

Figure 1.8. Northern autumn evening sky as seen from a relatively dark location. Stars shown to magnitude 5.0.

N

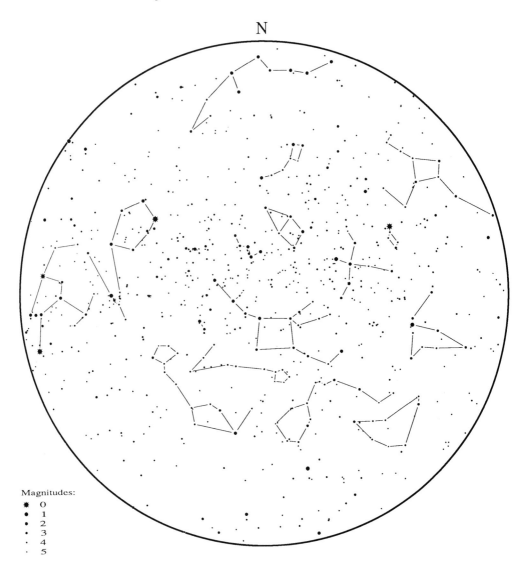

Magnitudes:
0
1
2
3
4
5

Figure 1.9. Southern summer evening sky as seen from a suburban location. Stars shown to magnitude 4.3.

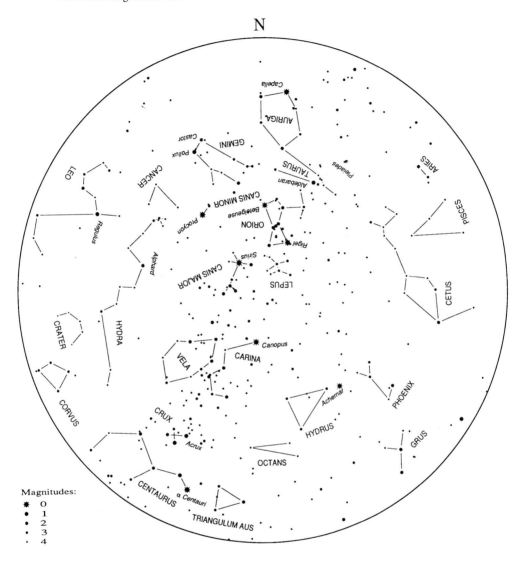

Figure 1.10. Southern autumn evening sky as seen from a suburban location. Stars shown to magnitude 4.3.

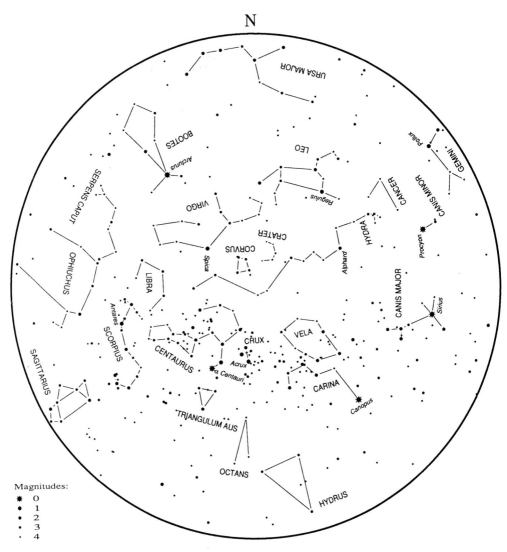

Magnitudes:
| | |
|---|---|
| ✳ | 0 |
| ● | 1 |
| ● | 2 |
| • | 3 |
| · | 4 |

Figure 1.11. Southern winter evening sky as seen from a suburban location. Stars shown to magnitude 4.3.

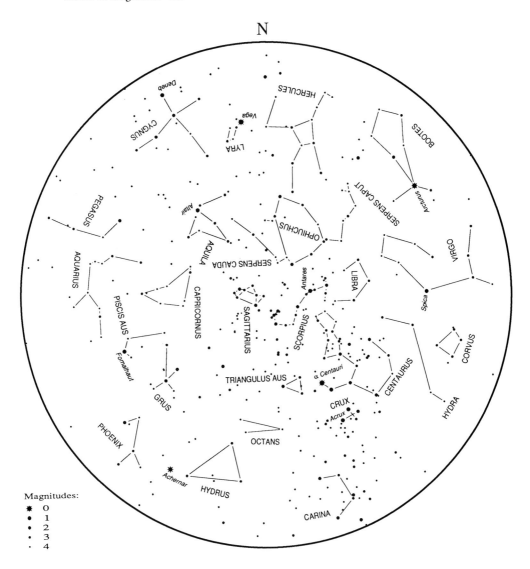

Figure 1.12. Southern spring evening sky as seen from a suburban location. Stars shown to magnitude 4.3.

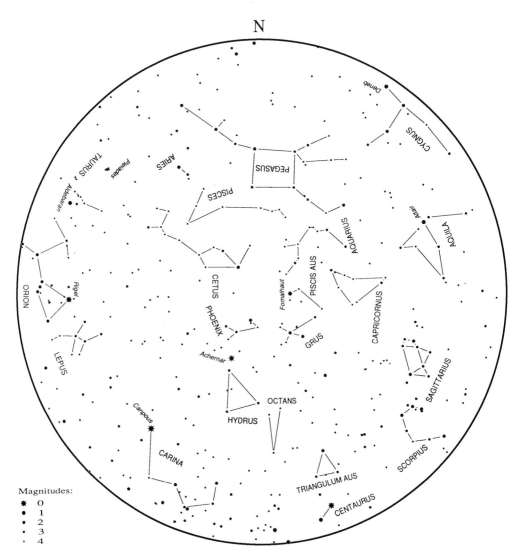

N

Magnitudes:
0
1
2
3
4

Figure 1.13. Southern summer evening sky as seen from a relatively dark location. Stars shown to magnitude 5.0.

N

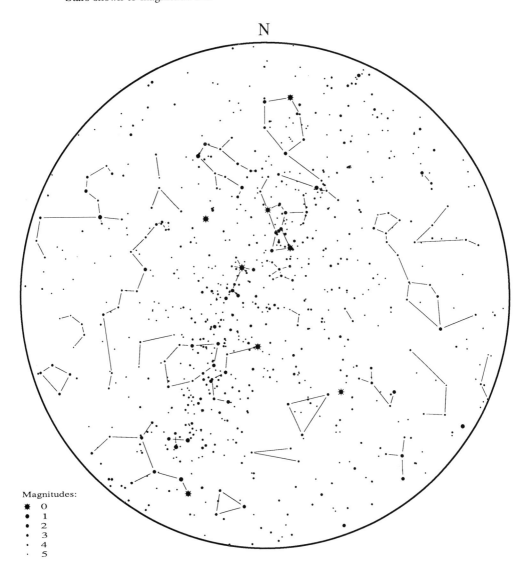

Magnitudes:

* 0
* 1
* 2
* 3
* 4
* 5

Figure 1.14. Southern autumn evening sky as seen from a relatively dark location. Stars shown to magnitude 5.0.

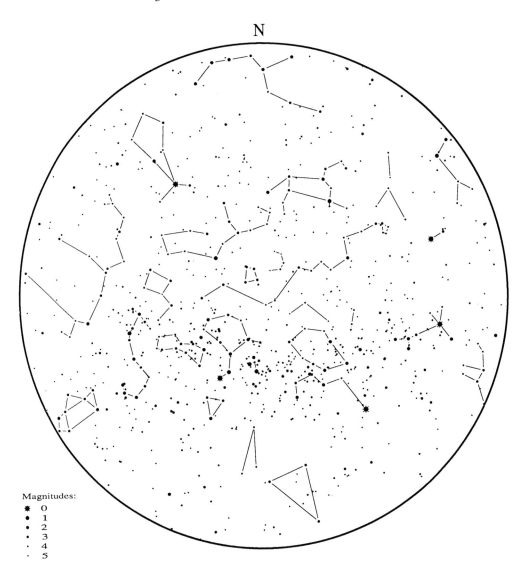

Figure 1.15. Southern winter evening sky as seen from a relatively dark location. Stars shown to magnitude 5.0.

N

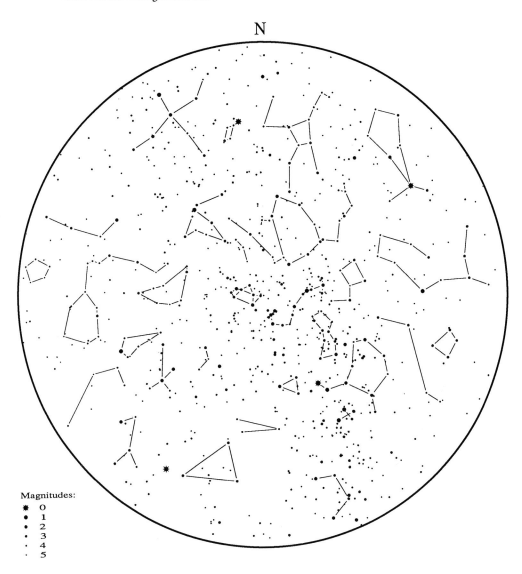

Magnitudes:
* 0
• 1
• 2
· 3
· 4
· 5

Figure 1.16. Southern spring evening sky as seen from a relatively dark location. Stars shown to magnitude 5.0.

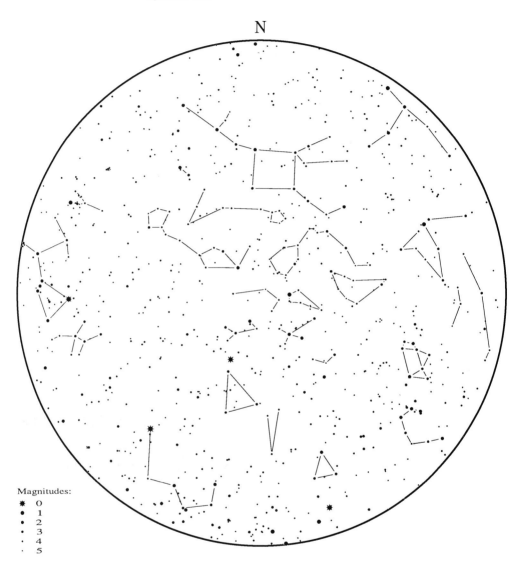

N

Magnitudes:
✸ 0
● 1
● 2
• 3
· 4
· 5

## 1.9   Starry, starry skies . . .

The words of this song, sung to the tune of the *Battle Hymn of the Republic*, were written by a prominent Canadian amateur astronomer.

> Mine eyes have seen the glory of the Whirlpool nebula,
> I have photographed M27 in Vulpecula.
> I know my way around the dust lanes of the Milky Way;
> I dread the break of day.
>
> Starry, starry skies in April,
> Starry, starry skies in April,
> Starry, starry skies in April,
> And clear warm nights in May.
>
> The wonders of the heavens are displayed each night to me,
> And Nature's laws reveal themselves in everything I see.
> I feel the vastness of the void, I hear each evening
> The constellations sing.
>
> Starry, starry skies in April,
> Starry, starry skies in April,
> Starry, starry skies in April,
> And clear warm nights in May.

Peter Jedicke, 1979

# 2   Without a telescope

Naked eye astronomy is an idea people laugh at. With all of the telescopes available, is there anything to do without a telescope?

The naked eye is actually a fabulous observing tool. It can capture tremendous contrasts, from bright sunshine to at least sixth magnitude stars at night. Coupled to a telescope, it can see details at the very limit of visibility on Jupiter, capturing at the same time the bright belts and zones that dominate the planet. It can detect faint comets and faint galaxies right next to glaringly bright stars.

With all these abilities, you might wonder why people bother with other detectors at all, like photographic film and electronic systems. The main reason is that the eye cannot be objective. Because it is connected to a brain, the combination opens a Pandora's box of subjective interpretations based on what you expect to see. Unlike film, which presumably records identical images when all the conditions are the same, the eye–brain combination will produce varying interpretations with each observation.

Whatever the eye's problems, however, it is still a great device to observe with. You do not need a telescope, or even binoculars, to begin your journey to the stars. As John Masefield suggested in his 1902 poem 'Sea Fever', all you need is 'a tall ship, and a star to steer her by' – the ship is your observing program, and the star is already there.

Some of the activities this chapter will suggest will be covered more thoroughly in future chapters. My purpose in the next few pages is to get you going on projects that need neither binoculars nor telescope.

## 2.1    Lights

### 2.1.1    Haloes

Some of the most beautiful sights in the sky are produced in our own atmosphere: the play of sunlight on clouds, flashes of lightning, rainbows, the effects of sunlight refracted through clouds. At night, we can see meteors in the upper atmosphere, but that is the subject of the next chapter.

*Solar and lunar haloes* are rings around the Sun and Moon that happen when their light is refracted through ice crystals in high cirrus clouds. When the daytime sky is filled with thick cirrus clouds, patches of light called sundogs or parhelia might form. Sometimes the cirrus blocks so much of the Sun's light that the real Sun and the sundog are the same brightness, and sometimes a pillar of light forms as well.

Haloes are a meteorological effect involving clouds only a few miles above us. Incidentally, the word meteorology means the science of the atmosphere, and the word meteor used to refer to any atmospheric phenomenon. Now it

Figure 2.1. Sundog photo by Dan Ward.

refers to the event of a collision between a tiny mass in space and the atmosphere. The subject of the next chapter, these events occur between 80 and 50 miles above us. Higher still, and in two zones around the planet's magnetic poles, the aurora occurs.

## 2.1.2   Aurora borealis and australis

July 8, 1966. With ten or fifteen young boys at the Adirondack Science Camp, I was setting up for a night of observing. The plan involved a group observing session for an hour or two, followed by a night of observing in blissful solitude.

But the twilight glow in the northwest refused to fade away, and when the glow began to shift from the west to the north I really was puzzled. Then slowly, out of that amorphous glow, a sharply focused ray of light appeared, brightened, and climbed the sky. Other rays then appeared until a curtain of green light shimmered before us. As our group stared silently at the magnificent display of graceful hues, we realized that this would be the night to forget all about telescopes, planets, and far-away galaxies. This night the show would be produced at home.

When charged particles from sunspots interact with the Earth's ionosphere, the result is a display of light that ranges from an amorphous glow to complex arcs of dancing rays. These eerie lights are most strongly visible from sites near the Earth's north magnetic pole as *northern lights* or *aurora borealis*, or around the Earth's south magnetic pole as *southern lights* or *aurora australis*. Since the magnetic poles are not that close to the geographic poles, people at similar latitudes do not have the same chance of seeing the aurora. Calgary, Alberta, and London, England, for example, are about the same latitude, but Calgary is very much closer to the north magnetic pole. Thus, while London gets frequent auroral displays, those over Calgary occur so often that they interfere with other forms of observing.

Aurorae can take many forms, the simplest of which is the greenish glow which resembles a haze near the horizon. With increasing strength the glow turns into a homogeneous arc, while rays gracefully towering from the arc turn it into a 'rayed arc'. Forms that look like curtains also can appear. Sometimes a 'corona' of rays appears directly overhead. Finally, a lot of rapid campfire-like movement is known as 'flames'.

Back in 1966, visual observations of aurora, or northern lights as they are more commonly known, were accepted eagerly by both the National Research Council in Canada and by the Auroral Data Center in the US. Although funding for these projects has stopped, the aurora has not, and it is fun to observe. Since the aurora is the effect of solar magnetic activity, observing these strange lights can ultimately tell us about our Sun as well as about our own upper atmosphere and their interaction.

The National Oceanic and Atmospheric Administration (NOAA) is still interested in reports of major displays seen through the United States, and in any display at all seen from the southern part of the country.

Figure 2.2. Aurora report, described in the text.

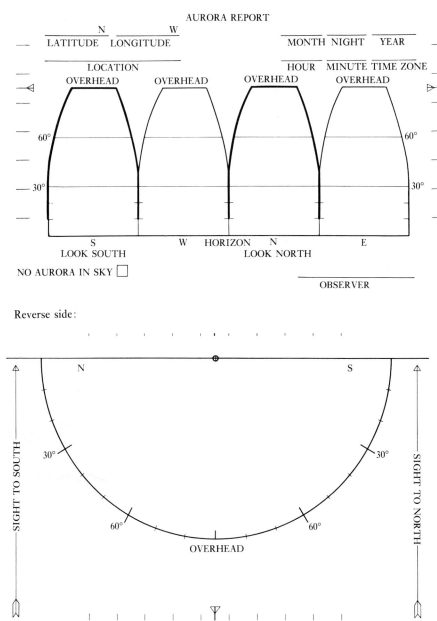

Figure 2.2 is an aurora report form which allows you to sketch the aurora on four panels that represent the sky's four directions. It offers a simple exercise in drawing what you see in the sky, an excellent prelude to the later telescopic observations of planets and other objects that you might try.

The idea is to make reports each night, and several times each night if there is a display. Look north first, plotting the aurora at its estimated zenith distance, the number of degrees from the point in the sky directly overhead. Because of the magnetic pole's location, a small display is more likely in the northwestern sky in eastern North America; such a display would appear more in the northeast, west of the Rockies. A major display could cover the sky. The four quadrants have each been divided into three parts for ease in sketching on this report.

The form's lower part is designed for use as a simple protractor to determine the distance in degrees from the zenith of an auroral structure. Attach a string with a small weight to the cross in the middle of the form. Then sight along one of the arrows toward the feature. The zenith distance in degrees is read where the string meets the protractor.

### 2.1.3   Zodiacal light and Gegenschein

The *zodiacal light* is a triangular-shaped glow that resembles a softly lit tepee. Always symmetric about the ecliptic, or plane of the Earth's orbit extended into space, this light is caused by sunlight reflecting off a huge collection of primeval solar system dust that has been replenished by recent comets. When the ecliptic rises high in the sky in winter evenings in the

Figure 2.3. The aurora, photographed by Stephen J. Edberg. This June, 1986, photo was taken with a 17 mm f/3.5 lens, 30 s exposure, on 3M1000 film.

northern hemisphere, so does the zodiacal light; in summer evenings, it is harder to see.

This dust cloud actually extends across the sky, joining the eastern and western zodiacal lights in a narrow *zodiacal band* which is extremely difficult to see, even under a very dark sky. However, this band widens into an oval spot, many degrees wide, that can be seen on a good night. This *Gegenschein* or counterglow, is centered on the ecliptic directly opposite the Sun. It is often so large and faint that you won't be convinced you have seen it. By averting your gaze, or looking out of the 'corner' of your eye, so that the rods in the center of your vision are not concentrating on this dim glow, you might have a better chance of detecting it. This technique is known as 'averted vision'.

### 2.1.4    Artificial satellites

Launched mostly by the Soviet Union and the United States, but by other countries as well, artificial satellites seem to be everywhere. They are most prominent in the first and last hours of the night, when they emerge from the Earth's shadow. Although the Soviet satellites tend to orbit south–north or north–south, and the American ones west–east, that rule of thumb is not dependable enough to determine the satellite's parent country with certainty.

It is sometimes possible to obtain visibility schedules for the US Space Shuttle when it is orbiting; when these times are favorable they are often broadcast over television. Do not expect, however, to see the geosynchronous satellite that bounces your TV signal without a telescope; it probably is shining at about 12th magnitude.

## 2.2    The planets

It is in the grand style of the pre-telescopic Arab, Greek and Roman observers that we should begin observing the planets. These observers pictured the five planets as wanderers whose paths through eternity took them across the sky. Because these objects were bright, and because they moved against the 'fixed' background of stars, they must have attracted much attention. Their wanderings are not aimless, only complicated, for each planet 'dances' to two drummers: its own orbit around the Sun, and its place relative to Earth. The planetary motions we see are a result of a combination of these two concepts. Watching the outer planets move is a fascinating pastime.

For a start, try recording from one week to another, either on paper or on film, the positions of whatever planets happen to be in the sky. If two or more planets are in the sky at the same time, you then can have even more fun noting the changes. If you record these motions on paper, all you need to do is draw just two or three of the constellations in which the planet of interest lies. Then, once each week, plot the planet's position against the background stars. If you decide to try photography, use a fast film and normal lens, and an exposure of at least five minutes. If you take a series of weekly photographs of exactly the same group of stars, by the end of your planet's viewing season you will have a record of its motion.

You will discover an interesting fact about how planets move. As a planet reaches its period of best visibility and is in the sky most of the night, it is said to be approaching *opposition* since it is near the opposite part of the sky from the Sun. At this time a strange thing happens. The planet's eastward motion among the stars will slow and then stop. Then the planet will appear to 'back up' for a while, just as if it forgot something! The planet has entered retrograde motion, an effect which you probably don't even notice every time you pass a car going your direction on a highway. At the moment you pass a slower car observe how the other car appears to move backward relative to yourself. Much the same thing happens on the solar system's 'highway'; as the Earth overtakes any of the 'superior' planets (planets further from the Sun than us; i.e. Mars through Pluto) these planets appear to move retrograde. The inferior planets (Mercury and Venus) never display the retrograde loops made by the outer planets but they do move about the

Figure 2.4. On January 28, 1984, Arizonian Tim Hunter caught five planets in a single exposure! Left to right, Mercury is very close to the horizon; Venus and Jupiter are close together; the Moon is not far from the star Antares; Mars, and Saturn are much higher; and the star Spica completes the line-up.

Sun in easterly, then westerly, directions. (Through a telescope these inferior planets also show moon-like phases.) These are truly dances to watch and enjoy.

## 2.3    Diversity of the stars

When looking at the stars, the first thing a careful observer will notice is the variety that we find in nature on Earth. They are not all the same brightness, nor are they the same color, nor are they distributed evenly over the sky.

In a 1967 book called *Beyond the Observatory*, the great astronomer Harlow Shapley made the impressive point that from a single glimmer of starlight, we can actually make at least 30 deductions! When you meet another person, you can make use of sight, sound, maybe even touch. Even with all that, it takes time to get to know that other person, even from a distance of a few feet. For a star, all we have is a single point of light, a point with no real size.

The first and most basic fact of what we could learn, said Shapley, was a star's position in the sky. Imagine a sky without Mizar, the middle star in the Dipper's handle – the Dipper would not be a dipper. The most casual look skyward will show that the stars are not distributed evenly over the sky. Close to the Milky Way, the plane of our galaxy, their numbers increase greatly. There are places where their density is very low. At areas far from the Milky Way, we are looking out of the galaxy into deeper space, so there are fewer stars.

A second deduction about the star is its brightness, which we have expressed as *magnitude*. Differences in brightness are obvious to see, and the fainter one looks, the more stars are seen. Stars are of different magnitudes for two reasons: their differing distances from us, and how bright they really are. One could draw an analogy to similarly constructed lampposts on a street, where the more distant lights appear dimmer than the closer ones. However, down the next street, the much brighter lights at a shopping center provide far more illumination even though they are farther away. A star's *apparent magnitude* is the result of how bright it really is, combined with how far it is away.

To get an idea of how intrinsically bright each star might be, imagine a sky in which every star is the same distance from us, say 10 parsecs or 33 light years. (A light year is about 5.8 trillion miles. If we measure a star that is 33 light years away while Earth is at one point in its orbit and then again six months later, it will show an angular motion of 10 arc seconds. Thus we can use the 'parsec', or parallax-second, as a unit of distance.) At that distance the Sun would shine dimly at magnitude 4.7. The summer triangle would appear very odd. Vega would shine at 0.6, just a bit fainter than we see it now. Altair would be a faint 2.3. Deneb would blaze at −7.2, several magnitudes brighter than Venus! These values are called *absolute magnitudes*.

Since they disregard distance, they provide a way of describing a star's actual brightness.

*Color* is another difference. An almost casual look will show that the brightest stars are not all white. Arcturus has an orange hue; Spica is bluish. Betelgeuse and Rigel in Orion show similar differences. These colors signify different temperatures, the blues being much hotter than the reds. What about yellow stars like the Sun? (Incidentally, the Sun is too bright to discern its color by eye.) Capella, the brightest star in Auriga, is such a yellow star, but much larger in size. The next of Shapley's deductions considers color more scientifically, examined using a *spectroscope* which divides a star's light in the same way that raindrops divide the Sun's light into a rainbow.

The different colors signify the various *spectral types* the stars have. Red Antares in Scorpius is known as an M star, on the cool end of the temperature spectrum, and Vega is a B star, at the spectrum's hot end. Our Sun (G2) and Capella (G8) are variations of the mid-range of G stars. The sexist mnemonic 'Oh be a fine girl kiss me' has served for years to identify the main spectral classes O, B, A, F, G, K and M. Adding the command 'Right now, sweetheart' identifies the rare redder classes R, N and S.

As Shapley's deductions continue and get more complex, they leave the simple studies that we have just made. But their point has already been made: starlight can teach us much about what lies overhead.

## 2.4   The Sun

**Warning: Never look at the Sun without proper protection!** (See Chapter 8.)

An easy and safe way of observing the Sun is by the *camera obscura* method. Punch a small hole in a large box. Cover this hole with foil, and cut a pinhole in the foil. On the inside of the box's other side, mount a sheet of white paper. Now turn the box upside down so that the pinhole is behind your shoulder. Face away from the Sun. The Sun's rays will shine through the pinhole and form a large image on the white paper. Remember not to look through the pinhole at the Sun!

With a correct sun filter (Welder's glass no. 14 is appropriate) you may see occasional dark markings known as sunspots. These spots can exceed the size of the Earth and thus be visible to the *protected* naked eye. Because the numbers of spots ebb and wane in roughly an 11-year cycle, there are long periods of time during which no naked eye spots are visible.

Observing the Sun in this way takes only a minute or two. Because the Sun will be without spots visible to the naked eye on many days, seeing one appear on one of the Sun's limbs, or edges, will be a special treat. It will be an even more special treat to watch the spot move across the Sun's face over a period of about two weeks as the Sun's rotation carries it along.

Figure 2.5 is a place for recording what you see on the Sun each day. It is designed so that a single sheet will last a month.

Figure 2.5. Report form for naked eye sunspots.

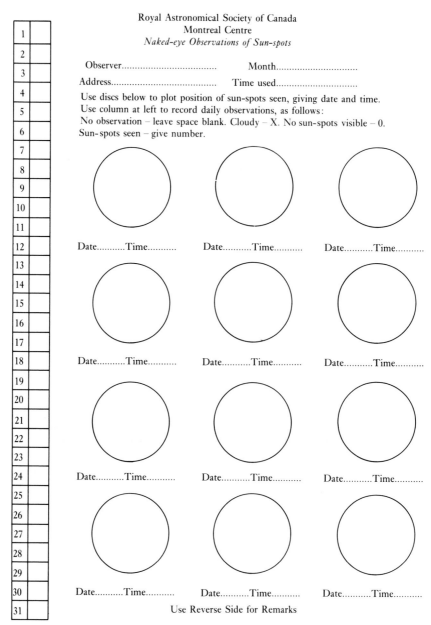

Royal Astronomical Society of Canada
Montreal Centre
*Naked-eye Observations of Sun-spots*

Observer.................................     Month.............................

Address......................................     Time used.............................

Use discs below to plot position of sun-spots seen, giving date and time.
Use column at left to record daily observations, as follows:
No observation – leave space blank. Cloudy – X. No sun-spots visible – 0.
Sun-spots seen – give number.

| 1 | |
|---|---|
| 2 | |
| 3 | |
| 4 | |
| 5 | |
| 6 | |
| 7 | |
| 8 | |
| 9 | |
| 10 | |
| 11 | |
| 12 | |
| 13 | |
| 14 | |
| 15 | |
| 16 | |
| 17 | |
| 18 | |
| 19 | |
| 20 | |
| 21 | |
| 22 | |
| 23 | |
| 24 | |
| 25 | |
| 26 | |
| 27 | |
| 28 | |
| 29 | |
| 30 | |
| 31 | |

Date...........Time...........     Date...........Time...........     Date...........Time...........

Date...........Time...........     Date...........Time...........     Date...........Time...........

Date...........Time...........     Date...........Time...........     Date...........Time...........

Date...........Time...........     Date...........Time...........     Date...........Time...........

Use Reverse Side for Remarks

## 2.5    The Moon

It is surprising how little is generally realized about the phases of the Moon. For instance, the queen of the night is actually in the sky *all* night only for a few nights each month, when it is near full phase. At first quarter it is in the sky at dusk and sets about halfway through the night, while at last quarter it does not rise until about midnight.

Another misconception is that full Moon is the best time to view our satellite. Wrong. As on Earth, shadows make lunar features stand out more clearly. When you can see the sunrise or sunset line (known as the terminator) on the Moon, the objects on that line will stand out clearly.

With the naked eye, how many features can you see on the Moon? At full phase, several large and dark plains (known as maria) dominate the topography, and as the phase changes, different aspects of these features become visible. Drawing the Moon's changing face from one phase to another will illustrate that point.

## 2.6    Mercury

Always close to the Sun, this innermost planet appears briefly as an 'evening star' for possibly two weeks, and then as a morning object for the same length of time. When Mercury is visible from Earth, it is considered to be at *elongation* from the Sun, meaning that it is relatively well separated from the Sun. When it is on a line of sight between Earth and Sun, it is at *conjunction*. However, even when Mercury is at its greatest angular distance it might still be near the horizon at sunrise or sunset, a geometry known as an unfavorable elongation. If the planet is far enough from the horizon at sunset to be visible for an hour or so afterwards, the elongation is considered favorable, and it is in such times that this elusive planet can be seen with the naked eye. The Royal Astronomical Society of Canada (RASC) *Observer's Handbook* as well as *Sky and Telescope* and *Astronomy* magazines give Mercury's viewing schedule.

## 2.7    Planets in daylight

Seeing planets, and even some bright stars, in daytime can be an interesting project. If you have seen Venus as an 'evening star' and know about where it is relative to the Sun, try finding it just before sunset by aligning with landmarks scouted the evening before. Without first finding it in a telescope, I have seen Venus, naked eye, at high noon, but my observing was through a deep blue sky of a mountaintop. Although Jupiter is a much fainter planet than Venus, it too is possible to see when the Sun is in the sky.

A small cardboard tube, like those used for paper rolls, will increase the contrast between the object you are seeking and the surrounding sky. With such a tube, finding daylight objects is easier. It will also enable you to identify fainter stars – perhaps by as much as a full magnitude – at night.

## 2.8    Variable stars

Now we return to our simple examination of the stars. Knowing that they have different colors, and different magnitudes, a natural extension of our observations of stars' brightnesses is to wonder if some of them change. We see a star brighten as it rises from the horizon, its light shining through less and less of Earth's atmosphere until it reaches the meridian, its highest point. Then the star fades somewhat as it sets. Also, a star will appear to twinkle as the result of the natural scintillation of the Earth's atmosphere. (The concepts of scintillation and seeing are discussed in Chapter 9.)

Besides these Earth-based changes, some stars actually do change their brightnesses, either because of the effects of a companion orbiting in front, or an actual change in a star's energy output. Later this book will devote an entire chapter to these *variable stars*, but for now let's look at some bright examples.

In the constellation of Cepheus is an interesting variable star called Delta Cephei. Discovered to vary by John Goodricke in 1784, it cycles through almost a full magnitude in a period of several days; at its maximum it shines at 3.5, then drops to 4.4 at minimum. Why not try observing this star each night, comparing it to nearby Zeta Cephei, whose brightness it matches at maximum, and Epsilon Cephei, which approximates Delta's magnitude at minimum? You don't even have to record magnitudes. If Delta is as bright as Zeta, assign it a value of 1. If it is slightly fainter than Zeta, give it 2; if it is halfway between Zeta and Epsilon, call it 3; if it is just slightly brighter than Epsilon, note it as 4; if it is as faint as Epsilon, write 5; if slightly fainter than Epsilon, mark 6.

Algol is another variable worth watching with the unaided eye. For a few hours over a period of slightly less than three days, this star drops in brightness from second to third magnitude as its companion star passes in front of it. Although Geminiano Montanari first reported Algol's change in 1667, it was not until 1782 that John Goodricke firmly established its identity as a variable star with repeated drops in magnitude.

Although Algol is eclipsed each three days, the times do not often coincide with evening hours. Some of these eclipses occur late at night or during daylight. Thus, your chance of casually catching a minimum are not that good. The RASC *Observer's Handbook* and *Sky and Telescope* include a list of future minima to watch for.

## 2.9    Deep sky objects

Because the nebulae, star clusters and galaxies appear more distinct against a dark sky background, they are usually relegated to telescopic observing under a good sky, as will be discussed in Chapter 17. However, a few can be seen under a city sky.

You might, for example, see the Pleiades, also known as the Seven Sisters, without binoculars or telescope. It is uncertain why they are known as the 'seven' sisters, as only six of the stars are clearly visible to the unaided eye. Perhaps in ancient Grecian times, one of the other stars was a half magnitude brighter; it would have taken only that much to brighten it to easy naked eye visibility.

It is fun to examine this cluster to see how many Pleiads we can see. Most observers see six. Since several stars are just a bit fainter than those, observers with keener eyesight or more experience often report not seven but eight, or even ten Pleiads.

Another naked eye cluster, at least as seen from a dark site, is M44, in Cancer. Because its stars are fainter, if you see the cluster at all, it will appear as an amorphous fuzzy patch of light, perhaps with a slightly mottled look.

Of the nebulae, or hydrogen clouds, the Great Nebula in the sword of Orion is the most prominent and is a naked eye object. From a suburban sky it might be seen as a misty spot; under a dark country sky it is clearly visible.

Because of their general faintness and low surface brightnesses, a factor explained in Section 1.1, galaxies are the most challenging type of object to see with the naked eye. However, under a dark sky you should be able to see Messier 31, the Andromeda Galaxy, as a misty spot. If you look carefully you should see its elongated shape. At some two million light years distance, this galaxy is the farthest object normally visible to the naked eye, the correct answer to an interesting trivia question!

From my site southeast of Tucson, I normally can see an even fainter galaxy, at about the same distance as M31. Messier 33 is in the constellation of Triangulum. Although this object is more challenging because of its extremely low surface brightness, it should appear under very dark sky conditions without any optical aid. For both these galaxies, a technique called *averted vision* might increase your chances of detection. Simply look out of the 'corner' of your eye, looking in another direction yet concentrating on the galaxy. Since the rods at your visual periphery are actually more sensitive than those at the center, they tend to reveal fainter things. Looking this way is rather uncomfortable, though effective for short periods of time.

## 2.10  Searching

Even in recent times, both comets and novae (stars undergoing eruptions) have been discovered with the unaided eye. Comet White–Ortiz–Bolelli, a bright comet making a hairpin turn about the Sun, was found that way in 1970, and five years later Nova Cygni was found independently by many observers simply by looking up. One of the three discoverers of the supernova of 1987 (see Chapter 16), Oscar Duhalde first observed it with his unaided eyes.

A beginner discovering an object will have a very hard time reporting it before more experienced observers do, but do not be discouraged. While navigating a Boeing 707 jet toward Oregon, an airline pilot named Stewart Wilson did discover and report Comet Wilson–Hubbard in 1961.

More than anything else, regular patrol work requires alertness, persistence, and knowledge. Besides, whether you are the first to actually report something should not matter when you are just starting out. To be able to discover a nova or comet by yourself, before you hear about it on the news, would be a significant achievement.

I begin each night with a look over the entire sky just to make sure nothing obviously unusual has happened. One night I found Nova Cygni of 1975 shining brightly at magnitude 1.6. At first I thought it was a very slow-moving orbiting satellite, and it was a few moments before my mind recognized that it was not moving at all and was a nova. Even though my find was a full day late, it was a thrill to get a personal introduction to the 'new star'.

Walter Scott Houston, this century's best known deep sky observer, was peacefully enjoying his after dinner pipe when Clifford Simpson asked about a bright comet to the west. Thinking of Comet Arend–Roland which had been bright a few weeks earlier, but had now faded, Houston answered, without looking up, that there was no comet. Simpson persisted; Houston looked, and saw a brilliant comet near the western horizon. Racing to his telescope, he started observing it at once before it set. His position at the telescope was unwieldy, so he put the pipe in his pocket, held on to the tube with one hand, adjusted the focus with the other, and began studying what he later learned was the new Comet Mrkos.

In a minute they smelled smoke. 'Scotty, you're on fire!' Simpson yelled. Houston threw the pipe and his burning jacket onto the grass, which also ignited. Finally, they smothered the flames.

# 3    Meteors

Now lies the Earth all Danae to the stars,
And all thy heart lies open unto me.
Now slides the silent meteor on, and leaves
A shining furrow, as thy thoughts in me.

Tennyson, 'The Princess'

Have you ever caught a falling star, as Perry Como's 1950s song asked? Maybe a falling star plummeted toward Earth one evening, and its fall attracted your attention and awakened you to the night sky.

That happened to me one July 4, about 1957; our summer camp in Vermont had finished celebrating Independence Day, and the sky was filled with stars. I noticed a particular star, probably Vega, high overhead. Just then a faint meteor appeared, moved across a small area of sky for an instant, then was gone. Indeed, the sky had not lost a star, but it did leave a seed in my mind that would grow years later.

For a romantic view of the sky, meteors are hard to beat. Even though we tell our children the difference between a star and a meteor, the fact that the two disparate things are linked at all is quite interesting. After all a star is a huge nuclear furnace perhaps a million miles across and many light years away, while a meteor is not a thing at all but an event taking place 60 miles above us as a particle of dust heats our atmosphere and disintegrates. The amateur meteor observer enjoys a kind of observing that is somewhat romantic too; this field does not require a dome or a telescope or an eyepiece.

The actual particle in space, typically no larger than a grain of sand, is called a *meteoroid*. As it enters the atmosphere, the event of its heating the atmosphere is called a *meteor*. If it is large enough to survive its fall and actually hit Earth, we call the rock a *meteorite*.

Meteor observing is fun whether done alone or in a group. It requires just a recorder and a timepiece of some sort, a clear night, and some knowledge of constellations and directions in the sky. Since a single observer cannot cover the sky, groups of two to eight can assemble for a sky-watching session that can last several hours, form lasting friendships, and still accomplish some good scientific data gathering.

The Delta Aquarids of July and the Perseids of August are ideal excuses for meteor 'parties'. Although two of the best meteor showers occur in winter, the Geminids of December and the Quadrantids of January, northern hemisphere winter conditions cause many observers to miss these fine celestial events. Those who have tried observing these winter showers have not been disappointed. It is believed that the record holder for the longest meteor observing session is George Alcock, the great amateur astronomer of England, who observed Quadrantids on the night of January 3/4, 1952, for 13 hours and 40 minutes.

## 3.1   Showers

In addition to the planets, asteroids, and comets, streams of meteoroids orbit the Sun too. These tiny particles in most cases have been shown to be remnants of comets or asteroids. The Eta Aquarids and the Orionid streams, for instance, share the same orbit as Comet Halley so it is logical to expect that Comet Halley is the 'parent' of these streams.

Comets contain different amounts of dust. A relatively dust-free comet would not be expected to be trailed by much of a meteor stream. In 1862, Periodic Comet Swift–Tuttle presented quite a show, with frequent eruptions of dust jets from its nucleus. This comet is the parent of the Perseid stream.

We do not see all the streams that comets produce. In order for a meteor stream to appear as a shower here, the Earth has to cross the orbit of the stream.

Meteor showers occur as Earth passes through these large streams of particles that travel in a comet's path. Shower meteors are traceable back to points in the sky called *radiants*. If you are observing on August 12, when the Perseid meteors reach their maximum you will likely see several meteors each hour whose paths, when reversed and extended, converge to a point in the constellation of Perseus. This is actually an effect of perspective. When Earth crosses a meteor stream where all the meteors travel in parallel orbits, the meteors obviously come from the same direction like the two rails of a railroad track. If you look down a railway track, you will see the rails appear to converge. Meteor showers offer the same effect. The place in the sky where the celestial rails converge is called the radiant.

From a dark country sky, you can observe meteors any night of the year. On any dark night you may see up to 20 meteors per hour, most of these coming from minor showers. These showers are older streams that are very well spread out, so that we are in their paths for several weeks. Some late November nights, for example, may be unusually rich, for even though no major showers are around, several less prominent ones, including the Leonids, Taurids, and Monocerotids, offer rich and varied meteor activity.

Nights on which major showers are near maximum can be very productive, offering you opportunities to compare the strength, duration, and radiants of the different meteor swarms by which the Earth passes. You will find that the showers do not just vary in numbers. Meteors from some showers are much faster than those of other showers, with a dramatic difference from the very slow Tau Herculids of June to the speeding Leonids of November.

The meteor rates in the following descriptions are called zenithal hourly rates, meaning that they have been corrected for the ideal situation where a single observer watches when a radiant is at the zenith. Since radiants are rarely that well placed, the rates you see will usually be slightly lower.

### 3.1.1   Showers month by month

**January**

*Quadrantids* Theoretically this is as powerful a shower as the August Perseids. Reaching its maximum on January 3, it can produce a good show of 50 meteors per hour from a radiant east of Alkaid, or Eta Ursae Majoris in the

Big Dipper. However, the maximum strength lasts only two or three hours. Unless the Moon is out of the night sky, and unless the maximum occurs at night and not during daylight, this shower could be quite weak, or missed altogether. The meteors are relatively fast, hitting the Earth's atmosphere at 41 kilometers per second (km/s).

## February

*Aurigids* A weak shower that peaks on February 9 from a radiant near Capella.

*Delta Leonids* A weak shower, detectable from early February to mid-March, and peaking on February 26. You will be lucky to see five meteors per hour at maximum. They have moderate speed, about 23 km/s.

## April

*Sigma Leonids* Peaking on April 17, these slow meteors (20 km/s) fall over a very long time span, from March 23 to May 13.

*Lyrids* Strongly variable in strength, these meteors average about 15 per hour per observer on the night of maximum, April 22. Their radiant lies near the bright star Vega. About three days after maximum, watch for possible bright fireballs. Their speed is relatively fast at 48 km/s. The Lyrids can be a surprising shower. In 1982 Earth passed through a dense part of the stream, during which one observer recorded a rate of 80 per hour.

## May

*Eta Aquarids* This is one of the two annual showers believed to derive from Halley's Comet. Although the shower is active for the month between April 21 until May 25, it reaches a peak just before dawn on May 5. They hit us at 65 km/s, appearing as fast streaks, some of which leave afterglows known as trains.

## June

*Tau Herculids* Reaching maximum on June 3, these meteors are detectable for about two weeks on either side of that date. Their speed is a slow 15 km/s.

*June Draconids* This weak shower peaks around June 28. The meteors radiate from a point east of Mizar in the Big Dipper, and are derived from Periodic Comet Pons–Winnecke.

## July

*Southern Delta Aquarids* A very fine shower, peaking on July 29 at a moderate 20 per hour rate. Their speed is relatively fast, 41 km/s, and their radiant is north of the bright star Fomalhaut. The shower's maximum night is made even more exciting by the presence of many early Perseids. The Moon can be the enemy of many a shower, for moonlight blocks faint meteors, severely limiting what an observer can see. The July–August meteor pageant is not as badly affected, since the Delta Aquarids and Perseids are separated by about two weeks. If one shower is hurt by the Moon, the other would not be. Only if the Moon is at first quarter for one and last quarter for the other would observations be impeded for both.

*Alpha Capricornids* Peaking on July 30, the meteors fall at an average rate of five meteors per hour, with a speed of 23 km/s. They are active roughly from July 15 to August 10.

## August

*Southern Iota Aquarids* The first of the August showers peaks on the 5th, with a maximum rate of five meteors per hour. Their speed is 34 km/s. They are active from mid-July through to the end of August.

*Northern Delta Aquarids* Peaking the same night as the Perseids, August 12, this weak shower averages two to five meteors per hour. Their speed is about the same as the southern Delta Aquarids, 42 km/s.

*Perseids* A wonderful spectacle! From year to year the apparent strength varies; some years I have seen Perseids coming in rapid succession followed by less active periods. Sometimes bright Perseids will appear to shatter; occasionally two or more meteors appear almost simultaneously, flying off in slightly different directions. Some very bright Perseids have cast shadows.

In a limited and theoretical sense, this is a period that should be avoided by all but the most experienced observers! If you are trying to understand from which shower a meteor belongs, Perseid time is very difficult since several showers are active at the same time and it is sometimes difficult to tell which is which. True, the others are somewhat esoteric streams, but so many of them are active at once that it makes the sky a busy and varied place. Nevertheless, all that activity combined with warm weather in the northern hemisphere make this period the best of the year.

*Kappa Cygnids* These meteors are visible from the second week in August to early October; they peak August 18; their speed is 26 km/s.

*Northern Iota Aquarids* Peaking August 20, these meteors have a speed of 31 km/s, a tiny bit slower than their southern counterparts. (You will not notice the difference visually.) They are active from mid-July to September 20.

## September

*Southern Piscids* This shower peaks September 20, although it is active all through September and October. Their speed is 26 km/s.

## October

*Annual Andromedids* Although these meteors peak around October 3, they are detectable from September 25 to November 12. Their speed ranges from 18 to 23 km/s.

*Draconids* Although this shower is not detectable every year, on October 8, 1946, a meteor 'storm' resulted when our planet crossed the path of debris from Periodic Comet Giacobini–Zinner, which had passed that point in space just a few days before. For five hours on that memorable night observers in Montreal counted over 2000 meteors, radiating from the head of Draco, and since they observed only through specially made rings that allowed viewing of selected areas, their total count was far below what the actual number of meteors would have been.

*Northern Piscids* Active from September 25 to October 19, these meteors peak on October 12. Their speed is 29 km/s.

*Orionids* Related to Halley's comet, this rich shower has a rate of 25 per hour; with a concentration of fireballs about three days after maximum. Like the Eta Aquarids, these are fast meteors, travelling at 66 km/s. They appear to radiate from a point northeast of the bright red star Betelgeuse.

## November

*Southern Taurids* This shower produces many fireballs during its run from late September to early December. Although it peaks around October 31, when its radiant lies south of the Pleiades cluster, its bright fireballs last through much of November. Often these fireballs will leave trains of ionization that last for a few minutes. The 1988 shower I saw produced many shadow-casting fireballs, one of which left a train for 15 minutes! Another one was so startling that my neighbor's dog barked in response. Their speed is 28 km/s, which is rather slow. These meteors derive from the famous Periodic Comet Encke.

*Northern Taurids* Peaking November 13 and lasting from September 19 to the end of November, these meteors are also slow at 29 km/s.

*Leonids* A good shower, at 71 km/s the fastest speed of any of the major streams. Peaking on November 17 from a radiant near the head of Leo, the normal maximum rate is 15 per hour. Every 33 years, however, the Earth crosses a dense part of the stream's orbit, around the position of the parent comet Tempel–Tuttle, 1866 I. If Earth's orbit intersects that of the stream at the right time, a storm often results.

On November 17, 1966, a group of dedicated observers on the east coast even rented a plane to carry them over the clouds in hopes of seeing something; after all, in the previous two years the rates had increased dramatically. As the sky brightened, meteors were appearing every minute, with the rate increasing rapidly.

In the southwest, where it was dark at the hour of maximum, the sky was blazing with meteors at a rate of 40 meteors per second! At any time at the height of the storm, the sky was filled with bright meteors.

In the late nineteenth century, November had two potential meteor storms. The *Andromedid* shower (different from the Annual Andromedids) produced big storms on November 27, 1872 and 1885. Those meteors came from Periodic Comet Biela, which had split into two pieces several returns earlier. Members of that stream have not been seen in many years.

### December

*Monocerotids* This normally weak shower put on a surprisingly strong performance in 1988. It peaks on December 10 but can be detected from November 27 to the end of December. Speed 42 km/s.

*Geminids* The largest shower of the year, boasting up to 75 meteors per hour. Radiating from a point near Castor, their speed is a moderate 35 km/s. If you ever get to see a bright Geminid meteor moving across the sky, you will know how beautiful a 'fireball' can be. Although the shower is within a quarter strength of maximum about 2.6 days on either side of the December 14 maximum, it peaks sharply and is best observed on the 14th between midnight and dawn.

Like the Perseids, this shower occasionally shows a clumping effect, with sometimes no meteors for up to five minutes, followed by a burst of four or five meteors within a minute. Thus, the 62 meteors I saw in a 75-minute period in 1987 appeared in such peaks that at times the sky seemed to be drizzling meteors!

In 1983 an object was discovered that shares the orbit of this stream. It is possible asteroid 3200 Phaethon is actually a comet that is no longer active, and is the parent of the Geminid meteors.

*Ursids* December's second major shower peaks on the 23rd from a radiant northwest of Kochab, or Beta Ursae Minoris in the Little Dipper. Its rate varies widely, but averages five meteors per hour. Their speed, 34 km/s, is about the same as that of the Geminids.

*Coma Berenicids* With no definable peak, this shower lasts approximately from December 12 to January 23. At 65 km/s, its meteors are fast-moving.

## 3.2    Observing procedure

### 3.2.1    Single observer

You will always see more meteors if you observe from a dark site, and for best results try looking at a part of sky about 45 degrees from the radiant and about 45 degrees above the horizon.

If all you want to record is a simple hourly rate, just use a mechanical counter that you press every time you see a meteor. At the end of an hour you have a record of how many meteors you saw. With two or more counters you can ambitiously record the numbers of various showers. Since I have only one counter, I use it to record the numbers of shower meteors and record the much lower numbers of nonshower meteors separately. On nights of major showers, you will get hourly rates this simple way, but not other important information.

With a tape recorder (and a mechanical counter as backup), you can record much more. The advantage of the tape recorder is that you don't ever need to look away from the sky. If you note the time when you start the tape, you can copy the data from the tape the following day, setting a clock to the time you began observing and recover the times that you observed meteors. For each meteor, record the following data:

**Magnitude** Record simply to the nearest half magnitude, comparing with stars such as those shown in Table 3.1.

With a little experience you will find many other stars to assist with magnitude estimates. We will discuss magnitude estimates in Chapters 14 and 16. A special problem with meteor magnitudes, however, is that since meteors appear suddenly, the factor of surprise tends to make some observers overestimate their magnitudes.

**Shower** By mentally tracing a meteor back, and remembering the speeds at which meteors travel, you can assign many of the meteors you see to a specific shower. If you cannot determine the shower, simply say 'nonshower'.

You should have no trouble determining the shower if you keep in mind the analogy of a railroad track and perspective. Tracing an imaginary path back to a likely radiant is an easy process which becomes second nature after a little practice.

**Remarks** Mention anything unusual. Was the meteor unusually fast or slow? Did it leave a post-meteor train, and if it did, how long did it last? If the meteor was bright, did it explode or shatter into pieces? Did it show an unusual color, perhaps red or green?

Table 3.1. *Comparison stars for meteor observing*

| Magnitude | |
|---|---|
| −1.5 | Sirius |
| 0 | Vega, Capella |
| 0.5 | Procyon |
| 1.0 | Spica, Altair, Aldebaran |
| 1.4 | Regulus |
| 2.0 | Alpha Ursae Majoris |
| 2.4 | Gamma Leonis, Gamma Ursae Majoris |
| 2.5 | Alpha Pegasi |
| 3.0 | Epsilon Geminorum, Gamma Ursae Minoris |
| 3.4 | Beta Bootis |

### 3.2.2    Group observing

Meteor observing can be fun for various sizes of groups. In a team of five, one observer takes north, another south, a third east, a fourth west, and a fifth acts as recorder. Observe together in hour-long periods, and allow time for breaks.

Organized by astronomy clubs, larger groups can divide the work in different ways. A good way of observing with as many as 15 people is to set up a set of four groups for observing and one for recording. A schedule would be prepared that would allow each observer one hour on, followed by half an hour off. Thus a two hour session might work as in Table 3.2, where each letter stands for a specific observer:

Anyone seeing a meteor would call 'Time!' The recorder then assigns a number, and asks for the magnitude estimate, shower membership, and comments.

Table 3.2. *Meteor watch observers' schedule*

| Time | North | South | East | West |
|---|---|---|---|---|
| 2200 | A | D | G | J |
| | B | E | H | K |
| 2230 | B | E | H | K |
| | C | F | I | L |
| 2300 | C | F | I | L |
| | A | D | G | J |
| 2330 | A | D | G | J |
| | B | E | H | K |

Some of my fondest memories from the 1960s are the meteor shower parties that Isabel Williamson organized for amateurs in Montreal. Using the special forms provided from Canada's National Research Council and reproduced in this book (Figure 3.1), we carefully recorded time, magnitude, and shower membership for each of the several hundred meteors we saw. The 1966 Perseid shower was our best; under a black sky, we saw 906 meteors in six hours of observing.

Figure 3.1. An IGY visual meteor report, developed by Peter Millman of Canada's National Research Council.

STATION

I.G.Y. VISUAL METEOR REPORT

DATE

| TIME ON | OFF | POS. |
|---|---|---|
| | | |
| | | |
| | | |
| | | |

TIME

| NO | MAG | SH | REMARKS |
|---|---|---|---|
| 1 | | | |
| 2 | | | |
| 3 | | | |
| 4 | | | |
| 5 | | | |
| 6 | | | |
| 7 | | | |
| 8 | | | |
| 9 | | | |
| 10 | | | |
| 11 | | | |
| 12 | | | |
| 13 | | | |
| 14 | | | |
| 15 | | | |
| 16 | | | |
| 17 | | | |
| 18 | | | |
| 19 | | | |
| 20 | | | |
| 21 | | | |
| 22 | | | |
| 23 | | | |
| 24 | | | |
| 25 | | | |
| 26 | | | |
| 27 | | | |
| 28 | | | |
| 29 | | | |
| 30 | | | |

TIME

| NO | MAG | SH | REMARKS |
|---|---|---|---|
| 31 | | | |
| 32 | | | |
| 33 | | | |
| 34 | | | |
| 35 | | | |
| 36 | | | |
| 37 | | | |
| 38 | | | |
| 39 | | | |
| 40 | | | |
| 41 | | | |
| 42 | | | |
| 43 | | | |
| 44 | | | |
| 45 | | | |
| 46 | | | |
| 47 | | | |
| 48 | | | |
| 49 | | | |
| 50 | | | |
| 51 | | | |
| 52 | | | |
| 53 | | | |
| 54 | | | |
| 55 | | | |
| 56 | | | |
| 57 | | | |
| 58 | | | |
| 59 | | | |
| 60 | | | |
| 61 | | | |
| 62 | | | |
| 63 | | | |
| 64 | | | |
| 65 | | | |

TIME

| NO | MAG | SH | REMARKS |
|---|---|---|---|
| 66 | | | |
| 67 | | | |
| 68 | | | |
| 69 | | | |
| 70 | | | |
| 71 | | | |
| 72 | | | |
| 73 | | | |
| 74 | | | |
| 75 | | | |
| 76 | | | |
| 77 | | | |
| 78 | | | |
| 79 | | | |
| 80 | | | |
| 81 | | | |
| 82 | | | |
| 83 | | | |
| 84 | | | |
| 85 | | | |
| 86 | | | |
| 87 | | | |
| 88 | | | |
| 89 | | | |
| 90 | | | |
| 91 | | | |
| 92 | | | |
| 93 | | | |
| 94 | | | |
| 95 | | | |
| 96 | | | |
| 97 | | | |
| 98 | | | |
| 99 | | | |
| 100 | | | |

TIME KEEPER

OBSERVER

Even though the atmosphere of fun resembled that of a party, the organization was militarily precise. Eight garden chairs were set up, two for each quadrant. One would be set up so that the observer who sat in it would gaze at the section of quadrant about 45 degrees altitude, and the other arranged so that its observer could watch the area below. Eight observers would be on duty and four relieved at a given time. Observers on break would be expected to rest inside, drinking coffee and otherwise resting their eyes for 20 minutes, and then go outdoors for the final 10 minutes to readapt their eyes for night watching. A team of three recorders, two on duty and one off, would keep time and record the meteors seen by all observers.

Instead of lawn chairs, observers in Ottawa used a permanently mounted system of boxes called 'coffins' into which observers could slip to retain body heat as they braved the frigid Canadian winter nights.

### 3.2.3    Hints

Serious meteor observing depends on keeping the sky conditions as consistent as possible. Although in an ideal world this means perfectly clear, meteor showers need observations under whatever conditions are available. It is important that you record these conditions frequently, especially if the sky has variable amounts of cloud. A thin cirrus covering that is evenly distributed over the sky actually is a very different statistical problem in reducing meteor counts from a condition involving heavy cumulus clouds that leave parts of the sky completely clear. With either type of cloud, you could get a rough idea of the shower's strength, but hardly an accurate one.

Be comfortable. An adjustable lawn chair makes a highly satisfactory observing accessory for meteors. You can set it up in advance so your head is facing the desired part of sky, and you can watch for an hour or so without straining your neck. The only disadvantage is that the chair may be so comfortable that you fall asleep! A portable radio or some other music might help, although the music will get transferred to your tape recorder along with your data!

With a standard or wide-angle lens, meteor photography can be fruitful. Keep the camera open for an hour or more. Normally a meteor must be at least $-2$ magnitude before it will record on film.

## 3.3    Fireballs

In September 1968 I was stunned by a bright flash in the sky behind me. It was a fireball at magnitude $-8$! Eight minutes later there was a low rumbling sound that possibly came from the event.

Any meteor brighter than about $-4$ (or the brightness of Venus, or bright enough to cast a shadow) is called a *fireball*. A meteor of this brightness or greater has a small chance of surviving its plunge through the atmosphere, and thus an accurate observation would be useful. There is a copy of a

Figure 3.2. An IGY fireball report, developed by Peter Millman of Canada's National Research Council.

**FIREBALL REPORT**

```
YEAR      MONTH      DAY
                              OBSERVER
          am
          pm
HOUR      MINUTE    TIME  ZONE   ADDRESS

WEATHER

LOCATION
OF
OBSERVER
WHEN                             BURSTS
FIREBALL
SEEN
                                 LUMINOSITY

                                 COLOUR

     LAT        LONG             FORM

                                 DURATION

                                 SOUNDS
                                              BEGIN
                                 POSITION
                                  IN  SKY     END

                                          ELEVATION      BEARING

                                 DATE              RELIABILITY

ACM - 1                          PLACE             REPORTER
```

fireball report form in Figure 3.2. Copy it and record carefully the path, the magnitude, the direction of flight, sounds, and any other interesting things for a fireball you see.

For their tiny size, meteors are a challenging field. From the rare experience of a meteor storm to the sight of a bright fireball, from the organized shower observation to the simple act of watching a meteor fall out of the sky, meteor observing is something special. Go out tonight, shower or not. Take a walk and see if a meteor accompanies you. See if two or three might identify a shower. When you watch meteors, you become a part of our planet's lazy journey round the Sun, picking up particles as it goes.

# 4    Choosing a telescope

A telescope is never mine unless I have walked into it in the dark or tripped over it. At least a small telescope will yield if you bump into it; large professional instruments are completely unforgiving. One night during an exposure of Halley's Comet with the 1.5 meter reflector in the Catalina Mountains near Tucson, I lifted my head to check something and bumped it into the base of the tube. The pain was incredible! When a few minutes later the exposure was completed, I saw a perfectly guided comet image. The telescope had not even noticed the encounter!

I hope that your telescope meetings will be softer. While a telescope is a beautiful thing, it really is a tool, a communication link between observer and sky. If you choose it carefully, the result will be an enjoyable, useful, and long-lasting investment.

If you rush into buying a telescope, you'll get bruised. That was the conclusion of a conversation I had one afternoon at a telescope store whose owner introduced me to a family looking for a telescope. 'What would you recommend?' the owner asked me. Surprisingly, the answer was not simple, for both children and telescopes come in all types and sizes. Even for a child, the act of choosing a telescope is complex, not to be taken lightly. If there were but one best telescope, then there would be only one telescope.

Actually, there are only three basic kinds of telescopes: the refractor, the reflector, and the compound system. Each type of scope is best for a specific type of observing, regarding both the object being observed and the observer. It is likely that of these three, the reflector will work out best for you because it offers the biggest aperture for what you spend. In any case, the quality of the telescope you buy is much more important than its type.

Before you answer what, perhaps you need to examine why. Do you want a telescope just to look around, or do you plan to do great things with it? Are planets your strength, or do you want to take advantage of a dark sky and look for galaxies? Do you want a telescope for your living room, to look at as much as through? Or do you perhaps want to build one yourself?

In a sense, there are almost as many varieties and sizes of telescopes as there are types and sizes of objects to look at in the sky. There are long refractors with ordinary lenses for the planets. There are refractors with fluorite lenses for better views of the planets. You have seen small, short refractors at department stores, but have you also seen the small stubby reflectors that are good for beginners? You could look at the complex catadioptric systems that are much more expensive, but their makers imply that they are much more convenient for everything. Or we could examine the Dobsonians, mountings made efficiently of wood, that satisfy our urge to look at the galaxies and clusters if we use them in a dark sky. If history is our interest, we could buy an old transit instrument that points upward to look at stars crossing the meridian. If it is astrophotography, then we should look for a telescope that gathers lots of light and is superbly well mounted.

What about mounts? Can we make do with a simple, two-motion altitude and azimuth setup? Or do we need a mount that, with proper setting up, lets our telescope follow a star automatically? Should we invest in a 'German equatorial' or a 'fork' mounting? Do we need a motor drive that will cause the telescope to follow a star automatically as it crosses the sky?

You don't have to rush into a choice, and besides, there is another way. Why not join your local astronomy club, and get to use the telescopes that belong to its members? Amateur astronomers are friendly folk; you can usually find one who will be willing to share an instrument with a new discoverer of the stars. In fact, your astronomy club may have a telescope or two reserved for the use of its newer members. (Both *Sky and Telescope* and

*Astronomy* magazines periodically run lists of astronomy clubs and contacts.)

The astronomy club approach has another advantage; you get to meet other people with varying levels of commitment to the stars. Most clubs meet at least monthly, with a lecture on some aspect of astronomical theory or observation, and then they have a monthly 'star party' during which most of their telescopes are set up on a field, often away from city lights. At such a star party, you can get a pretty good idea of what telescopes are all about by looking through one after another. Chances are you will find one type that suits you.

This approach does have a disadvantage. Sometimes you may encounter a club person who assumes that you are more familiar with the field than you are and that either you understand all he has to say, or you're not bright enough. Most clubs have at least one person like this, so don't let that bother you; just find someone else, and enjoy the good teaching a club can offer you.

## 4.1   Binoculars

While you are deciding on a telescope, why not try a pair of binoculars? The universe was created so that people with binoculars would have something to do. One of the most versatile optical instruments ever devised to observe nature, binoculars are ideal for starting our day by watching birds, enriching it with a study of herons on a lake, a view of moving traffic on a bridge or a closeup of a critical baseball play, and ending it with a silent look at the stars. Each view offers the chance of discovery, of a new interest, of something about our environment, or of something genuinely new. Some of us may even have begun our love for the stars by a long look through a pair of binoculars.

A pair of inexpensive binoculars has a proven ability in astronomy. Even if you are just beginning your stellar journey, a first look through binoculars can tell you much about what the sky is made of. They offer the distinct advantage of making use of both your eyes, while most telescopes send their light only into one eye. Look at the Milky Way, and 'discover' that it is not a continuous line of light but an array of countless faint stars. Look at the planet Mars, or Jupiter, or Venus, and 'discover' how its position among the faint stars changes from night to night. Look carefully at small groups of stars within constellations, and find for yourself some brand new miniature patterns. Not far from Albireo, at the foot of the Northern Cross, is a curious little asterism that looks like a coathanger. A little faint for the unaided eye, it overfills the field of most telescopes: it is an ideal target for binoculars. Can you 'discover' any other intriguing shapes?

Although any set of binoculars will do for astronomy, I recommend a good quality set of 7 × 50s. The 50 mm diameter lenses, with eyepieces that produce a seven times magnification, are tailored for the opening of your eyes when they are peering into a dark sky and the pupils are wide open. Of course, they are also excellent for other purposes. The stars demand better quality optics and construction than do most other uses. If you want a pair of binoculars to look into your neighbor's window, you don't need to worry about whether the set is a 'field glass' consisting of two parallel sets of lenses, or a modern 'prism binocular' with a complex set of eyepieces, prisms and objective lenses. Either type will show the lamp and chair inside the window. Nor do you care whether your binoculars are German-style, with similar cover plates at each end of the body, or American, with a cover plate only at the eyepiece end and a somewhat more streamlined shape, if all you want to do is see what is happening next door. If people are moving quickly from window to window in your neighbor's house, you would likely want to buy a set that has a central knob that focuses both lenses at once, instead of one that needs to be focused individually, lest you miss something exciting. In fact, most central focus binoculars offer a correction focus on one of the eyepieces, to account for eye differences among different users. Your astronomical set may be so large that you need to mount it on a tripod in order to make the star images steadier. You would definitely not want such a pair for your ground-based spy project!

### 4.1.1   Anticipating problems

How can we tell if a pair of binoculars is of good quality? Price helps, but there is more to it. Although you can get binoculars almost anywhere, don't let their ubiquity fool you into thinking that buying a pair will be easy. Most of the cheaper sets, and some of the good ones, are flawed and shouldn't be in your observing inventory. The only way to find a good pair is to test several at a store.

Before you pay for a pair of binoculars, insist on testing them. Do the images at the edges of the field look fuzzy, even though what you see at the center is clear and in focus? If you notice this effect with distant buildings, trees or mountains, it will be much more obvious with star images.

How well does the pair focus? Does the image seem to stay roughly in focus (but not supersharp) through a half-turn or more of the focusing knob, or does it come to a very sharp focus only at one point of the turn? If it is the latter, your set has superior image quality.

Now try checking the alignment. The image should be identical in both glasses. Put the binoculars on a steady support, a tripod if possible, or at least brace them against a car or wall of a building. Looking at a distant object on the horizon, or if you don't have a horizon try a distant car license plate, first close one eye then the other. What is at the center in one lens should be at the center at the other. What is near the top, bottom, left, and right edges should be at the same place in the other lens. If the images match at the center but not at an edge, the eyepiece magnifications may not be well matched. If the optical systems are of different powers or are misaligned, their images will not appear as one to your eye, and you should not buy them. If the misalignment is slight, your brain will actually try to complete the alignment, with a headache as the likely result.

Usually, the more you pay for a pair of binoculars, the better aligned they will be. You still need to test them. I have rarely seen a cheap pair in perfect alignment, and I have found some very expensive glasses in poor alignment. You simply have to spend the time testing the binoculars at the store. If the salespeople won't let you, walk away.

There are numerous models of high powered binoculars (i.e. $20 \times 80$) on the market today. In fact, I have not seen any of these pairs in good alignment! Worse, most of these you buy through the mail, and the companies normally object when you try to return them. Never buy a pair of binoculars that way.

## 4.2 Telescopes

Even after you graduate from your simple binoculars, you should always keep them, for the sky is always receptive to the fine views they offer. The night will come when you realize that you are ready for your own private telescope. But such a variety of telescopes from which to choose! In all the stores, at all the astronomy clubs, everywhere you look, there seem to be more and more models of telescopes. Which is best?

Before you make a decision, examine the features of each type, and then you should be able to answer that question for yourself.

### 4.2.1 Refractor

This is the traditional telescope that *looks* like a telescope. The instruments that reach out the dome in comic strips, looking plaintively toward the sky, are refractors. These telescopes were the sky queens during the nineteenth century, and telescopes like the ones at Lowell and Yerkes Observatories, and at Meudon near Paris, still are used to do good astronomy.

The first known refractor was the result of a thought by Galileo of Padua, in 1609. Once his idea had come to fruition, he had a 'spyglass' that consisted of two lenses held apart at a special distance by a tube. When he looked through this refractor, he saw the heavens 'enlarged a hundred and a thousand times from what the wise men of all past ages had thought'.

A proud and bold man, Galileo eventually took his telescope for all its worth, and unfortunately paid a high price for it by having to retract all the discoveries he had claimed. Perhaps Galileo pushed his arrogance too far, and Pope Urban VIII reacted as a victim of a science he did not understand. But the simple optics of the small refractors Galileo made and used offered a bold concept about the nature of our place in the universe.

The modern refractor is an improvement of the telescope that Galileo used to discover that Jupiter had moons, that Venus had phases, and that the Sun had spots. The old instrument consisted of two lenses. At the top is a large lens called the objective. This lens gathered the light and bent it so that it reached a focus at the eyepiece, which was at the other end of the tube.

Such a telescope is deceptively simple. Its two lenses try to bend the light from a planet or star to a focus, but how can they, when different colors focus at slightly different distances? As observers began to understand this interesting property of colors of light, they tried to minimize it by making their telescopes very long, so that the color effect became less distracting. The longest refractor ever used in this way was an eighteenth century behemoth of 150 feet. Later this problem was tackled by changing the objective from a single lens to a pair of lenses held together. Made of crown and flint glasses, this combination lens would compensate for the different wavelengths of light, bringing them to a better focus. Other than this significant change, there have not been that many alterations in refractor technology over the years. The lenses got much better, especially with people like Alvan Clark and John Brashear grinding absolutely fabulous lenses during the nineteenth and early twentieth centuries. With the computerized processes available now, and a three-element lens called an 'apochromat', today's lenses are as good or better than any made in the past.

Refractors have served astronomers for almost 400 years, and they still offer fine performance if (this is a big 'if') they are well made and well mounted. Even though a toy (typically a refractor telescope selling for under $75) will outperform Galileo's telescope by quite a margin, such a telescope is not nearly good enough either for you or a child, because their views of the sky are poor and could destroy a budding interest in the stars. Although their optics are not very good, their main drawback is sloppy mountings that make it impossible to keep them steady for productive observing. If you wish to spend only that amount of money, try binoculars. Good refractor telescopes are expensive.

At any price, a long refractor is decidedly not the best telescope for looking at dim galaxies and other objects that require a dark sky. These objects have what we call low surface brightnesses, meaning that their light is spread out over a large area of a telescope's field of view, and not much brighter than the surrounding sky background, and thus a telescope with a wide field of view is required. A traditional long refractor cannot offer such a field, although short refractors with wider fields do well on these deep sky objects.

Long refractors are made for planets. Their narrow fields and good optics provide the needed contrast to make the subtle details on the surface of a distant planet stand out. You will be especially happy with how a good refractor shows you the Moon. The details of the craters, mountains, and valleys of the Moon, as well as the subtle marks within the belts of Jupiter's atmosphere, come out very well under a refractor's watchful eye.

The disadvantages of a long refractor are its size and expense. Because light has to travel completely through a lens, the quality of glass has to be superior to that in a mirror which just reflects light back. Also, a long tube is susceptible to a breeze which would shake it, resulting in an unsteady image.

### 4.2.2    Reflector

Reflectors use mirrors instead of lenses to form an image. In this type of telescope, light travels down the tube to be reflected by a curved mirror which forms the image. The light then travels back until it reaches a flat secondary mirror whose purpose is to direct the light to an eyepiece located conveniently at the side of the tube.

Newtonian reflectors, as these telescopes are known, have been popular ever since the second one, devised by Isaac Newton in 1671. Newton could not possibly have had any idea of how successful his design would become, still the most popular amateur telescope design three centuries later. In fact, his first instrument, built in 1668, was almost a secret; it was his second that aroused the interest of the Royal Society of Great Britain.

It is true that such reflectors are considerably less expensive than are refractors of the same size. This does not mean that they are not as good; in fact, Newtonian reflectors are more widely used by more experienced observers than any other type. Since all wavelengths of light reflect to the same focus, and since light simply bounces off a mirror rather than passes through it, special glass is not necessary. Modern mirrors are usually made of pyrex rather than plate glass because pyrex is not as sensitive to temperature changes that temporarily distort the mirror's figure.

### 4.2.3    Compound telescopes

Another type of reflector, known as the Schmidt–Cassegrain or catadioptric, has become very popular in the last 20 years. This type of telescope uses a correcting lens at the front as well as a mirror at the back to gather and focus light. Such telescopes send the light down the tube one extra time, and as a result they are more compact. They are also more expensive than Newtonians.

Which is best? Reflectors lack the compactness of the Schmidt–Cassegrain, and usually a long refractor offers a finer, more detailed image of a planet. But when one considers price and overall convenience, I recommend for a first telescope the reflector.

## 4.3    Eyepieces

Choosing eyepieces is as important as choosing the rest of our telescope, for again there are many types of eyepieces. Get the best eyepiece you can afford, for a good telescope without a good eyepiece is not a good telescope.

Kellner eyepieces have a design that allows for a reasonably wide field of view, which means that when you look through the eyepiece you will see a reasonable amount of sky. The trouble with older eyepieces, of designs called 'Ramsden' or 'Huygenian', is that their fields of view were so small that it was hard to find objects and keep them centered.

Erfle eyepieces are more expensive than Kellners, since they use more elements of glass to produce a much wider field of view. A modern variation is called a 'wide angle' or 'ultra-wide' eyepiece, which has a field of view about the same as an Erfle but which is designed in a way that image quality at the edge of the field is almost as sharp as at the center. (To improve edge-sharpness still further, some companies offer expensive 'coma correctors'.)

Orthoscopic eyepieces offer smaller fields but better quality than Kellners. These eyepieces are good for long-focus refractors, which do not need as much field correction as the short reflectors do.

Choosing an eyepiece may be harder than choosing a telescope; in a recent magazine I found over 300 different manufacturer–dealer combinations of eyepieces. The simplest advice is to get the best optics you can afford. If you are on a budget, the Kellner design, including the refined 'RKE', provides you reasonable quality, and some orthoscopics may also be available for under $75. As your tastes refine, you may later want to augment your instrument with more sophisticated eyepiece designs that use several elements of glass. Plossl and Nagler are two such designs.

## 4.4    Mounts

It is really unfortunate that mounts are often overlooked when one chooses a new telescope, because if the mount is weak, the telescope it supports will be useless. The mount should be sturdy but also not so heavy that you cannot move it. There are two basic types, of which the simplest is known as an *altazimuth*. It comes with two axes, one to move the telescope up and down (altitude or elevation), and the other to move the instrument across the sky in lines parallel to the horizon (azimuth). The *equatorial* mount differs in that these two axes are oriented so that one parallels the Earth's axis, aligned north–south in azimuth with the altitude equal to your latitude. Thus, one axis moves the telescope in declination while the other moves it in right ascension. These words describe the latitude and longitude of the sky, and a telescope so mounted can follow a star by moving one axis instead of two.

## 4.5    Why not make your own?

Here is another solution to the problem of choosing a good quality telescope if you do not want to spend a lot of money to get it. It does not necessarily mean that you need to grind your own mirror, for there are companies that sell completed mirrors at reasonable prices. Making your own telescope could involve several levels of ingenuity on your part:

1. Choosing separate tube assembly and mount. Companies occasionally sell completed tube assemblies; you could buy one and then make a simple mounting.

2. Buy a completed mirror and separate parts, and assemble them yourself. This way has some attractions. By choosing the parts yourself, you have ultimate control over what goes into your telescope, from the size and quality of its mirror to the type of supports you use for it and for the diagonal. You can design and build the mounting. However, going along this route presents quite a challenge, since you need to ensure that the parts fit together – optical, mechanical, everything – not a small problem! This challenging project will certainly give you respect for what has to go into a decent telescope. Although instructions for this are beyond this book's scope, so to speak, there are other books with good suggestions (see the bibliography in Chapter 24).

3. Grinding mirrors. The thrill of starting with a circular blank of flat glass, then grinding, polishing, and figuring it into an accurate optical device that gathers light, has attracted many people into amateur astronomy in the past. In fact, the study of astronomy and the building of telescopes were once much the same thing. Galileo and Herschel opened their eyes to the universe with telescopes they had made themselves. Around 1840 Leon Foucault developed a knife-edge test that is still used widely to check the figure of a mirror, and began the use of a second glass blank instead of a piece of iron as a tool to grind glass into an optical surface.

In the third decade of this century, two men began to spread their enthusiasm for grinding a mirror and constructing a reflector telescope. Russell Porter was an Arctic explorer, a painter and sketcher, and a telescope maker. In 1925 Albert Ingalls, a reporter from *Scientific American* magazine, met Porter and was enthralled with his dedication. His resulting article in the magazine's November 1925 issue marked the joining of the hand of Porter with the pen of Ingalls to turn the idea of telescope making into a popular and engaging avocation.

That November article in *Scientific American* helped launch an amateur telescope making (ATM) movement that prospered for more than 50 years. The movement is now fading, temporarily one hopes, because of the easy availability of inexpensive optics and the lack of easy availability of telescope making materials. Porter's designs, his art, his skill, and his enthusiasm infected thousands of ATMs over many years, not just to save money on their telescopes, but to experience the birth of a new telescope. His designs included a 'Springfield mount' whose eyepiece always stayed in the same place, as well as an ornamental 'garden telescope', a work of art in bronze with a six-inch mirror resting in a pattern of lotus leaves. By pointing the telescope toward the Sun and safely projecting the bright image, the instrument also became a sundial.

In 1926 the first annual meeting of Stellar Fane, Shrine to the Stars, took place at the summit of Breezy Hill, near Springfield, Vermont. Quickly shortened to Stellafane, that meeting has provided a safe haven for generations of 'TNs' (telescope nuts) and has grown very large, so that these days the convention attracts more than 2000 people. If you become more interested in telescopes, you will want to visit Stellafane, a gathering that gives you the chance to see the latest in unique telescope design. For western United States amateurs, each May on the American Memorial Day weekend, the Riverside Telescope Makers Conference meets near Big Bear Lake, California.

## 4.6    Extremes

This chapter opened with a comment about the many varieties of telescopes. Phoenix amateur astronomer Peter Manly has searched away from the main types to the odd and extreme, and has come up with some treasures such as mirrors made of strange volcanic glasses like obsidian.

The odd things he has found in the back room of the world's telescope collection include an all-sky camera made of a polished hub-cap, whose reflective surface sends light from the entire sky to a camera lens. There are telescopes mounted on trailers, both for amateur and professional use. Several observers have attached observing chairs to their telescopes.

The three largest refractors, Manly tells us, are the Yerkes Observatory 40 inch, still in use, and two that have not survived; the 41 inch made but never used for Pulkovo Observatory, and the Great Paris 49.2 inch of the 1900 World Exhibition. This telescope used a 'sidereostat' mirror 78 inches (2 m) in diameter that followed an object, reflecting its light into the horizontal telescope tube which rested in one position. A section of tube that included the eyepiece was mounted on a railway track. This strange telescope did not survive for long after the Paris exhibition, although the main lens is on display at the Paris Observatory.

Although the highest permanent observatory site is at 4305 m Mount Evans, Colorado, where a 24 inch reflector looks skyward through very thin air, not all telescopes are rooted to the ground. The Hubble Space Telescope is but one of several instruments designed for use off the ground. Telescopes mounted in balloons, and the Kuiper airborne observatory, result from our drive for a darker and clearer sky.

The Sun has different observational needs from anything else. It requires telescopes not only to gather light but also to take advantage of high magnification. Thus, there are two tall tower telescopes on Mt Wilson, and the 500-foot long McMath Solar Telescope at Kitt Peak designed for solar study.

Radio astronomy uses interesting designs for its telescopes that observe the sky in wavelengths other than optical. Grote Reber's back-yard 31-foot radio telescope was among the first. Bell Labs has a horn antenna. Curious in appearance, its work seemed routine until Arno Penzias and Robert

Wilson used it in their discovery of the background radiation probably left over from the 'Big Bang' of the universe's inception.

# 5     Recording your observations

Observing has two parts: observing and recording. You cannot do one without the other, for observations locked away only in your mind are not observations, and data do not really exist unless they are recorded somewhere. Early in your observing career you should develop the habit of recording what you see. Find a procedure that will support your own needs and interests. Some people use index cards, others use large ring binders, still others may use computers.

There are three purposes for keeping records: for yourself, for others, and for posterity. The first purpose gives you the opportunity to review and remember what you have observed. The second allows others to share your work, so that you prepare reports to submit to astronomical societies and even to professional astronomers. Future generations define the third purpose, something you likely don't worry much about now.

On November 27, 1872, the English poet Gerard Manley Hopkins observed an extraordinary event, which he recorded in his journal:

> Great fall of stars, identified with Biela's Comet. They radiated from Perseus or Andromeda and in falling, at least I noticed it of those falling at all southwards, took a pitch to the left halfway through their flight. The kitchen boys came running with a great todo to say something redhot had struck the meatsafe over the scullery door with a great noise and falling into the yard gone into several pieces. No authentic fragment was found but Br. Hostage saw marks of burning on the safe and the slightest of dints as if made by a soft body, so that if anything fell it was probably a body of gas, Fr. Perry thought. It did not appear easy to give any other explanation than a meteoric one. Br. Starkey saw and heard also but was odd and close about it.[1]

Because Biela's Comet had split, returned as two comets, then vanished, and years later returned as a meteor shower, it is still actively referred to as a splendid example of the death of a comet, and thus Hopkins's story is more than a historical footnote (see Chapter 14). Hopkins kept a detailed journal that included notes such as this on unusual natural events he saw. How could he possibly have thought that more than a century later, Periodic Comet Biela and its Andromedid meteor shower would still be a subject of interest? Had he not written down his impression of this night, it would have been lost.

For our first purpose, any sort of record that you understand is probably sufficient. A simple notebook with a diary-like description of what you have observed is fine for most of what you need to remember: what you saw, and

1 Gerard Manley Hopkins, *The Journals and Papers of Gerard Manley Hopkins*, eds. H. House and G. Storey, pp. 227, 8. Oxford University Press, 1959.

your reaction to seeing it. The second purpose requires that you describe enough of your experience that another person could reconstruct your observation and use the data you have obtained. Thus, your record should be legible, complete, and if you use codes, they should be properly explained. Something to keep in mind is that you never know when or how your records might be useful to someone later. Information about a bright fireball may be useful if later someone recovers a piece of the fall. A report of a bright northern lights display might be useful for someone trying to study the history of such events. Of course, if your record is well written and clearly understood, it will be satisfactory for posterity as well.

I use note books – 15 of them that in the last 25 years have recorded the details of over 7600 observing sessions. Here are a few examples:

6418M. Wed, Oct 5/6/ 0440–0530/ 10-f/ JM4/ Miranda/ –/ CN3 0h 30m east.

Translation: 'On the morning (M) of the night of October 5/6, I began session 6418 at 4:40 local time. The session ended at 5:30. The sky was superb, with no trace of haze or even cirrus clouds. I observed in station 4 of Jarnac Observatory, using my 40 cm f/5 reflector with standard 32 mm Erfle eyepiece, named 'Miranda'. I hunted unsuccessfully for comets in the eastern sky for a half hour'. (CN3 is the code name for my comet hunt project. When I started it in 1965, I had intended that novae may be searched for as well, and the name 'CN' has never been changed.)

7600EM. 1988 01 12/13/ 1840–0130/10-f/ JM3, 4/ Minerva (19 mm), Miranda/ Jim Scotti, Alister Ling/ Comet Liller 1988a (with Minerva): 1988 01 13.10 M1 = 9.9 DC = 1 Coma Diam = 5'. CN 3 1h 30mE.

In this session, I observed with two telescopes, Minerva, which I know is a 15 cm f/4 reflector, and Miranda. With me during this session were two other observers.

The first item of business was a magnitude estimate of a newly discovered comet, which I was able to find with the smaller telescope. Using stars whose magnitudes I knew, I estimated the brightness of the comet as 9.9. The comet had a degree of condensation of 1, which is almost completely diffuse with virtually no central condensation, and its coma diameter was 5 arc minutes.

7536E. Saturday, October 10/11 1987 /1900–2030/ 10-f/JM3/Pegasus/-/ CN3 1h15mW. Comet Bradfield 1987 10 11.12 M1 = 6.6. Tail 1.3 degrees at PA 90. ?Possible comet in Bootes.

In this session I estimated the brightness of Comet Bradfield, and noted a 1.3 degree long tail pointing due east. I was not too certain, but I also strongly suspected that I had discovered a new comet. The following night, during session 7537E, I would see the comet in a different place and record 'new comet confirmed, 14h 38m.3, +17 degrees 18'.' It turned out to be Comet Levy 1987y.

Records can note the presence of an unusual marking on Mars or even atmospheric effects like complex halos. They can note your first sighting of a new constellation, or an unusual experience you had while observing:

> Session 7585AN Dec 20/21/ 1730–0700/ 10-f/ Bigelow/ 61-inch, Minerva. [To reach the observatory] I left the car safely on the side of the road. Then I walked with a suitcase of winter clothes + briefcase in one hand AND Minerva, with mounting, in the other hand up the road to the 61, mostly through 6 to 8 inches of snow on ice. When I finally got to 61 I was absolutely exhausted. I observed lots of comets . . .

Notice how, by naming telescopes and otherwise coding information, one can cut the time and effort that is needed to keep good records. The 10 in the record is part of a personal code I devised many years ago that still works well in summarizing sky conditions:

0   rain or other precipitation
1   completely clouded sky
2   mostly clouded sky
3   partly clouded sky
4   very hazy sky
5   moderately hazy sky
6   slightly hazy sky
7   suburban clear sky
8   unusually good suburban sky; Milky Way in Cygnus visible with averted vision
9   good rural sky; Milky Way and M31 visible
10   superb country sky; no trace of haze or cirrus

One night, while observing at the rim of the Grand Canyon, I recorded 11. The faintest star I could see without optical aid was magnitude 7.3, which is at least 2.5 times fainter than the faintest star one normally can see without a telescope!

**What you could include** One of the most useful things you can do is to keep a set of notes of observing aspects you think of as you proceed with a session. These notes can range from your qualitative impressions on the sky conditions, how well your telescope is working, even your mood. You never know when one of these notes will come in useful later, and 'later' could go on for centuries. In the Observatoire de Paris is displayed one of the original notebooks of the eighteenth century observer Charles Messier. The opened page showed some Jupiter observations he had done, and it contained some thoughts on his optics. As I read his words the centuries melted away and I had the powerful feeling of being with the great comet discoverer as he set about arranging his telescope and eyepieces in anticipation of a fine clear night. From his page of notes I could imagine myself as his contemporary, and that his observing methods were not very different from mine. If in technical form we have the advantage of more efficient telescopes and electronics, the basic themes of observing and appreciating the night sky are the same, and forever locked in records like these.

# PART TWO

# Moon, Sun and planets

## 6 The Moon

Many people see the Moon as romance, and many astronomers see it as inconvenient. The astronomers need dark sky conditions to study many of the faint objects that demand the darkest possible sky background. In brightening up this background, the Moon has taken some of our precious dark sky time away from us.

I see the Moon as poetry, the way we were taught when we were young. Its phases elicit moods, its light shines off clouds to paint beautiful pictures of night, and its eclipses are wonderful. With the smallest telescope we can climb its mountains and explore its craters from our back yards. With huge plains called Serenity, Tranquility, Clouds (Mare Nubium), and Storms (Oceanus Procellarum), the Moon is indeed a poetic place.

The Moon has a lot to answer for. The subject of poetry and dreams, astro-dollars and politics, the Moon has inspired us for a long time. Not so many years go, Project Apollo made the Moon a central target for observers. Proposed landing sites were the objects of detailed drawings and photographs. Occultations of stars by the Moon were heavily observed, and in the late 1960s a program for observing lunar transient phenomena became very active. Now that spacefaring eyes are no longer turned in her direction, Lady Moon has lost some of her courtiers, and with today's emphasis on large, fast telescopes, her romantic appeal for amateur observers has dropped.

I think that is unfortunate. The Moon does justice to almost any small telescope. I cannot remember the number of nights I have turned my telescope towards the Moon for a productive hour of Moon gazing. The Moon has something to offer that no other celestial body can hope to match: it is so close to us that it really is a *place*. You can climb the mountains, explore the huge lava fields, walk along the Straight Wall, picnic in the shadow of Copernicus's central peak, and witness the conditions of a rising or setting Sun. The Moon is a geologist's dream, for all of its rocks are exposed for our study, uncovered by vegetation. It is a dead world only in the sense that its lack of atmosphere and extreme temperatures prevent life forms from flourishing. In scenery and beauty, the Moon is very much alive.

## 6.1    Why observe the Moon?

Is there anything at all that is left to do, or have the Rangers, Surveyors, Lunas, Lunar Orbiters, Lunokhods, and Apollos answered every possible question?

Observationally, you would be hard pressed to make any major new selenographic finds, except in some special areas. One, 'Luna Incognita', is an area that is rarely well seen from Earth and not at all by spacecraft. On occasion it can be seen with a small telescope, and visual studies of this area are always helpful.

The other areas are observationally hazy. It began in the 1950s when some observers occasionally reported flashes of light that could have been caused by impacts of meteorites on the Moon. To confirm these reports, the Association of Lunar and Planetary Observers sponsored a search for lunar meteors, a carefully timed program during which, for three selected nights each month, observers would watch the dark side of the Moon for a short period of time. This was a search for flashes, success depending on a flash being confirmed by another observer. Although the program no longer runs, its former popularity deserves at least this historical mention.

A more recent program has grown out of the Lunar Meteor Search, and, although the aims differ, the procedure is much the same. Called the search for lunar transient phenomena, observers try to note any temporary coloring or brightening on the Moon's surface, particularly on some selected features where activity has been reported. As with lunar meteors, the problem remains of having at least two good observers independently see a transient event at precisely the same time. There will be more about this in the next chapter.

Another program of historical interest was promoted by a group at Louisiana State University's Department of Ornithology. It involved the observing and recording of high-flying migratory birds as they appeared to fly in front of the Moon. The program was active for at least 15 years, from 1946 to 1960. Although it has nothing to do with astronomy, it does require a small telescope capable of seeing the entire face of the Moon in one field. One wonders why more people do not follow this program now during migratory seasons, for it does take advantage of a sky made almost uselessly bright by the full Moon. A prominent Montreal observer named DeLisle Garneau realized that migrating birds flying against the Sun could also be counted in the same way. After he saw his first, he commented that 'I knew it wasn't a sunspot when I saw it flap its wings.'

## 6.2   The phases

Questions like 'How much later does the Moon rise each night?' are not as simple to answer as they might seem. Although the average is some 50 minutes, the real answer depends on how far north or south the Moon is on its ecliptic path, on your latitude, and on where the Moon is in its elliptical orbit of the Earth. In the northern hemisphere, the rising Moon in late summer may come less than 15 minutes later than the night before, and on winter mornings may rise over an hour later each night.

These phases allow you to observe the Moon on its own terms, night after night. Part of these terms mean that we see essentially about one half of the Moon; the other, known as the 'far side', always points away from Earth. When a crater or mountain range is fully sunlit, there is little contrast to show it clearly. However, when the Sun rises or sets on the moon, it creates beautiful shadows that accentuate these features. The changing border between the light and dark sides of the Moon is called the terminator, and it provides the most dramatic views of anything on the Moon. You can follow the terminator as it marches across the Moon's face, taking you on a tour of the Moon's most exciting features.

## 6.3   Training project

On October 8, 1960, I first entered the Observatory of the Montreal Centre of the Royal Astronomical Society of Canada, intending to become an observer. Expecting many handouts with forms and charts, I was surprised to leave with only one – a lunar map showing over 300 features. 'Take this map home,' I was told; 'look through your telescope and view every one of the 300 craters and 26 other features that are plotted.'

I found this to be a most useful training exercise, the object of which was to make a copy of the lunar map with each crater plotted only after it has been observed. This exercise teaches care and discipline at the eyepiece, and because each crater has its own best observing time during the lunar month, the exercise also teaches planning. It actually took me over a year to complete the identification of all 326 lunar features. Figure 6.1 is a simplified version of the map I prepared in 1964.

I strongly suggest that you begin your serious observing with this program. The lunar map is still available from Sky Publishing Corporation, and it also appears in *Norton's Star Atlas*. And, given time, your own copy, drawn as you observe feature after feature, will appear in your observing log. The features are easiest to see when they are near the terminator. What follows are some informal notes on what you might expect along the terminator for each day of the 28-day cycle. As these notes include only a small part of what is worth seeing each night, they encourage you to make your own discoveries. They do, however, give you a guide for what to expect each night.

Figure 6.1. It is easy to spot the features named on this home-made chart, which is a simplified version of the lunar map I prepared as part of the training exercise completed in 1964 using some small telescopes and the *Sky and Telescope* lunar map described in the text. Each feature described in the day to day notes in this chapter is labeled.

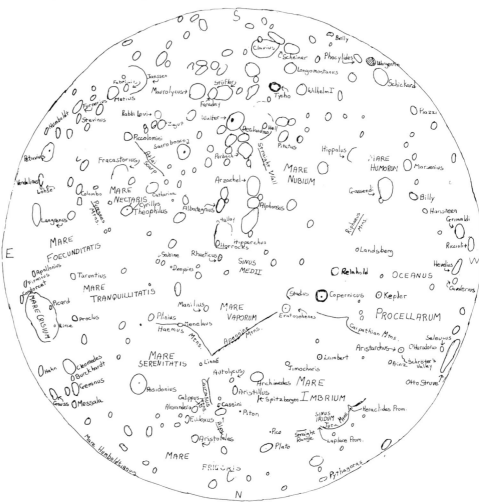

This tour will do more than show us lunar features; it will introduce us to ancient astronomers who made their mark long ago.

Before 1961, classical directions of east and west on the Moon referred to directions as we see them from Earth. With growing interest in the Moon as a body in space, the International Astronomical Union reversed those directions. Now, east is where the Sun rises on the Moon, while west is the direction of sunset. Older maps may show the classical directions. In the following description I use the IAU standards for east and west. North and south remain the same.

# 6.4  Day to day notes

*Day 1:* On some occasions it is possible to see a Moon less than a day old. On June 15, 1988, Brian Skiff and I were observing from Kitt Peak with a 6-inch reflector when we saw a tiny sliver of lunar crescent barely 17 hours old. When the new Moon is 'above' the Sun it is possible to see, for a few moments, a Moon less than 24 hours old. Studies of very young Moons are historically important, for civilizations using lunar calendars relied on early observations. With a Moon about a day old, a southern crater named Humboldt, and a northern mare called Humboldtianum, appear as small dark streaks. Because they are so close to the Moon's edge, as we on Earth see them, the craters appear very elongated.

*Day 2:* This day begins with our first view of the beautiful plain called Mare Crisium. (Although the word mare denotes a sea, in its modern lunar sense it means a basaltic plain.) You can enjoy Crisium's shadowed mountain borders as the Sun just begins to light up its eastern side. I remember an eclipse of the Moon our group watched in late December, 1964. The night was frigid but school had just ended for the holidays. As the Earth's shadow advanced to this great plain, someone called out 'Mare Crisium!' in a way that sounded so much like Merry Christmas that we had a good laugh. Langrenus and Petavius are the highlight of this evening's display. Petavius is a delight. With its beautiful and complex central mountain peaks, towering over 8000 feet above the crater floor, Petavius is one of the best sights the young Moon has to offer.

**Earthshine** This reflected light allows us to see the night side when the Moon is in a crescent phase. Dominating the Moon's sky, the Earth illuminates the landscape of the dark side, much more so than the light of full Moon lights up the hemisphere of Earth it is shining on. We see the Moon's night side lit up by this reflected light. Why not call it earthlight? If moonlight is light from the Moon shining here, then earthlight is light from Earth shining there. Earthshine is earthlight reflected back to us; moonshine would mean (among other things) moonlight reflected back to the Moon.

Earthshine adds to the Moon's beauty at crescent phases. It lights up the dark side softly and subtly, allowing us to explore its darkened features, and search for the strange things called lunar transient phenomena, which we will explore later.

*Day 3:* Where Mare Crisium was just becoming visible last night, now we can see it in almost all its glory. Can you see any ghost craters (old craters that have been partly covered by newer lava flows) on its floor? Crisium's southern neighbor Mare Fecunditatis is now making its debut. Although the Fecunditatis is considerably larger than Crisium, it is less conspicuous because it lacks the high mountain borders of Crisium. An extension of Fecunditatis passes just west of Vendelinus and Patavius. For the first time tonight we also got a good look at a chain of features, the large walled plain

Cleomedes, the high-walled crater Burckhardt, crater Geminus, and the harder to see Messala.

*Day 4:* Mare Crisium is now fully visible, and what an area it has turned out to be. Surrounded by a host of beautiful and different craters, it is truly a sight. If we 'drive' around the shores of this mare, we stop by crater Proclus (lying just outside the mare), and even take a look at Picard which stands out like an island in Crisium. Picard is named for a prominent seventeenth century observer, known particularly for his observations of Comet Halley in 1682, and for his suggestion that a great observatory in Paris be established. Toward the north is another 'island', the crater named Peirce. One of the oldest lunar features is prominent tonight; Janssen, a walled plain whose early history has been overwritten by more recent impacts. Janssen's walls have been shattered in several places, so that it no longer appears as a completely formed feature.

Mare Fecunditatis is also fully visible, and we now turn our attention to some of the other emerging maria. Mare Tranquilitatis, the Sea of Tranquility, and Mare Nectaris, are just beginning to yield to sunrise. We also get to see the sharply defined Taruntius, whose walls act as a border crossing between Fecunditatis and Tranquilitatis.

*Day 5:* The 'star' of this day is the huge crater Theophilus, easily found right where Mare Nectaris, fully visible tonight, meets Mare Tranquilitatis. Theophilus is one of the Moon's finest craters, displaying a high central peak consisting of several mountains. It forms an interesting crater pair with its neighbor Cyrillus, a crater whose boundary looks so square that it is hard to define it as a crater. South of Theophilus is the Atlai Scarp, just beginning its day as a curved fault line, bright against a darker surface background. It parallels the shore of Mare Nectaris so closely that it must have been formed at the same time. At the scarp's southeast end is Piccolomini, a sharp, deeply cut feature. Mare Nectaris is also easily visible tonight.

Fracastorius may have been a complete crater once, but today all that is left is an indentation on the south end of Mare Nectaris. Here is a place where a chapter of the Moon's evolution is written in clear language for us. Where is the north wall? When the lava flows that built Mare Nectaris reached the old crater, their forces rounded much of it down to some ridges and hills.

*Day 6:* As the Moon approaches its first quarter, Maurolycus is just beginning to show its large dimensions. Like Theophilus, this crater has a complex central peak. On this night, Mare Serenitatis is completely seen. The Haemus Mountains, acting as the southwest border of Serenitatis, are also beginning to appear, complete with Menelaus, a small but sharply defined crater. The Caucasus mountains are still mostly hidden, but two sharp craters on their border are visible, Eudoxus and Aristoteles. As the terminator reaches the center of the disk at first quarter Moon, the Moon offers you a choice, depending on something called libration. As the Moon orbits the Earth it slowly sways, or librates, so that sometimes we get to see regions in the far northern or southern regions that are otherwise lost to view. Because of this libration we actually see some 59% of the Moon's surface from Earth, although only 50% at any one time. The effects of this libration are visible over just a few days, and actually make an interesting observing project. As the Moon passes first quarter, note carefully the north and south limb features that are visible, and record how they change. As new features become visible on the north limb, those at the southern limb will disappear.

The project may interest you because the causes of libration are complex. Because the plane of the Moon's equator is inclined slightly to that of its orbit, we can occasionally see up to 6 degrees of the 'far side' of the Moon, either on the north or on the south limb. This libration in latitude is accompanied by one in longitude that allows a peak of about 7 degrees of the 'far side' of the east and west limbs. This east–west libration is most noticeable near the new and full phases. A third libration effect results from our own planet's rotation, and is most noticeable at full Moon. During our view of the Moon on such a night, our base of observations is moved by the Earth's rotation by a distance of some 7500 miles. We call this small change in the appearance of the full Moon the 'diurnal effect'.

*Day 7:* Although the Caucasus and Haemus Mountains are fully visible, the special crater to see tonight is Hipparchus, for it shows up well only on the night the terminator passes across its face. Inside its ancient walls are some newer craters, particularly Horrocks and Halley. Two plains have also appeared, the Mare Vaporum and the Sinus Medii as well as Albategnius, an oval walled plain. Sunrise has also hit Piton, the most prominent of the Moon's isolated peaks. Tonight is the best night to see its sharp peak right on the terminator.

*Day 8:* Perhaps the most interesting night of the lunar month, this is a night for mountains, craters, and the Straight Wall. The Alps and Alpine Valley are visible. The Moon's finest range, the Apennines, shows its full glory. Also the best known of the plains, Mare Imbrium, is becoming prominent as the rising sun exposes more of its surface. Archimedes is also in sunlight now, forming a beautiful triangle with two other craters, Aristillus, the larger one, and Autolycus. Nearby we can see the Spitzbergen Mountains. We can also get our first good look at Plato, a huge oval walled plain whose dark floor is always easily found. South of this huge crater the peak of the large mountain Pico lights up the terminator.

On this magnificent night we get our first good look at Alphonsus, whose central peak has been the object of some strange reports. This is also the night we see the Straight Wall, a ridge that is probably a fault line. It looks like a cliff with vertical drop, but actually the drop resembles more a steep slope of about 45 degrees. Nearby is crater Ptolemy.

Can you see Deslandres and Walter, a pair of features which represent events at different times in the history of that area? Deslandres is a huge walled plain, but a very old one whose eastern wall had been tromped on by the meteorite fall that formed Walter. On the eastern floor of Deslandres, near Walter's wall, is a strange brightening first seen by the seventeenth century Italian–French astronomer Giovanni Cassini and now known as Cassini's bright spot. If you don't see it very well tonight, it will become easier as the Sun shines more strongly on it in the next few nights. Probably related to Tycho's ray system, it is likely more recent than Walter. On the other side of Deslandres is a small crater we should look at, if only because it is named for an astronomer and clergyman named Hell.

*Day 9:* This is the night of Copernicus and Tycho, as both craters dominate the terminator. Both craters are fabulous sights throughout their two week day. When Tycho is near the terminator, the rays near it are not easily visible, although the ones farther away are seen. Named for the late sixteenth century astronomers who so radically changed our thinking about the universe, these craters honor Nicolaus Copernicus, the Polish scientist who devised the Sun-centered theory, and Tycho Brahe, whose observations of the supernova of 1572 called into serious question the unchanging nature of the 'sphere' of the stars.

Named for the astronomer of Cyrene who made, around 240 BC, the first observational estimate of the size of the Earth, the crater Eratosthenes is almost fully visible tonight. It is sharp and deep, although nearby Copernicus tends to draw our attention from it. North of Copernicus is a small but sharply defined crater called Pytheas, and further north still is the larger but less well defined Lambert. South of Tycho is the huge walled plane Clavius. This is a huge feature, its long side being some 150 miles wide. It is named for the Bavarian Jesuit astronomer Christopher Clavius, who in the late sixteenth century helped prepare the Gregorian calendar.

Notice how well Plato shows up this night, along with the Teneriffe Mountains that are near its border with Mare Imbrium. North of Plato is the Mare Frigoris, an often ignored plain which is larger in area and longer than any other observable mare except the Oceanus Procellarum.

*Day 10:* Tonight's highlights are with mountain and ridge peaks, rather than with craters. A group of mountains known as the Straight Range, as well as the Jura Mountains, the Sinus Iridium, and the Laplace and Heraclides Promontories, are all prominent tonight. Promontory is a good word; you can see how these features protrude into the Mare Imbrium. To the southwest of Copernicus, a small bright crater called Lansberg is coming into view, and even further to the southwest is the Riphaeus Range. Finally, west of Clavius and just receiving the first rays of Sun is Scheiner, a large walled plain.

*Day 11:* As the Moon continues to wax into a strongly gibbous phase, the rest of the sky is brightening up remarkably. However, the Moon's brightness is still less that *one half* of what it will be on the night of full phase! We are already losing our chance to observe deep sky objects, so we observe the Moon and notice that the features which looked so obvious just a few nights ago are now difficult to see as they are now fully lit by sunlight.

This is a night of maria, and the largest of the lunar plains, the Oceanus Procellarum, is showing more of itself with each night. Mare Fecunditatis, Mare Undarum, and Mare Spumans are also visible if the longitude libration is favorable. Not far from the terminator is Kepler, now almost fully bathed in sunlight. Although it is only about 20 miles wide, it is a complex crater with a bright system of rays. Johannes Kepler, of course, was the great astronomer and mathematician who determined that the orbits of the planets follow ellipses, and the discoverer of the great supernova of 1604. North and west of Kepler is Aristarchus.

Named for Aristarchus of Samos, who estimated the relative sizes of the Earth, the Moon, and the Sun, Aristarchus is the brightest spot on the Moon, and like Alphonsus it is suspected of occasionally showing a dull red glow. Hidden in the glare of Aristarchus is Herodotus, a crater with a bright rim but not given its due because of its proximity to bright Aristarchus. North of Aristarchus lies Schroter's Valley. Resembling a snake, its southern end is known as Cobra's Head.

*Day 12:* By this day Gassendi is almost fully illuminated, and to its south we see Mare Humorum. Nearby is Schickard, a walled plain with a bright border. However, tonight is the night to start looking for the crater that almost touches Schickard, a strange crater called Wargentin, which looks for all the Moon like a crater that has simply been turned upside down. Known also as 'Thin Cheese', this pancake-like crater was probably normal at one time, until lava started coming up from a source inside the crater, and stopping just before it started to overflow, especially on its northern part. With no breaks in the crater's wall to let out the lava, the flow eventually hardened.

Incidentally, with only two days until full phase, the Moon's light is still only one half of what it will be.

*Day 13:* Tonight we should see Hevelius, along with its southern neighbor Grimaldi and its northern neighbor Cavalerius. (Francesco Maria Grimaldi, 1618–63, discovered the effect of diffraction of light around objects.) Though difficult to spot, crater Seleuchus has a long ray system. Pythagoras, a large walled plain but very much foreshortened so that it appears as an ellipse, is also worth a look tonight. Every high school mathematician knows of Pythagoras; in addition to his famous theorem, he also propounded the view that the Earth is a sphere.

*Day 14:* The full Moon dominates the sky tonight, so that not much else in the deep sky is visible. Also, few of the Moon's own features show easily except those at the limbs which are lit by the Sun at a sharp angle. The ray systems do appear best at full Moon, since they need direct light to be well seen. The rays from Tycho, Furnerius, and Stevinus cross each other in an intricate network which is best seen at this phase.

*Day 15:* We now begin to see the sun setting on all the features we watched as the Lunar day began. As the sunset terminator begins to take over the east limb, we often forget that the west limb is still in early morning, so that features on both limbs can be seen. On the east side we can watch Mare Crisium gradually go into darkness, as well as the walled plain Gauss. Also, the Mare Fecunditatis is well lit for good study on this night. Humboldt, the large walled plain we observed two weeks ago, is getting dark.

Mare Humboldtianium shows a different perspective now that the Sun is setting on it. The crater Hahn looks particularly striking at sunset. Theophilus is showing a very different structure from what it displayed at sunrise.

*Day 16:* The dark surface of Mare Crisium is beginning to command our attention once more as the Sun becomes low over it. The mountains at its borders are beginning to cast shadows. With its high central mountain, Petavius is striking. Furnerius, Langrenus, and Vendelinus all show exciting structure on their floors, although these details are difficult to see unless you catch the terminator within a critical period of a few hours. Compare the sharply walled Langrenus with the much softer (and older) Vendelinus.

*Day 17:* Langrenus is disappearing at sunset, along with Petavius, with its high central peak. As the terminator moves westward, Mare Crisium reveals more of its complex structure. We can see some craters north of Crisium, like Cleomedes, Burckhardt, Geminus, and Mesalla.

*Day 18:* Tonight we enjoy a host of beautiful features in the sharp shadows of sunset. Mare Crisium is mostly gone now, but instead we have several sharp craters like Taruntius, and some old, faded ones like Colombo. We get a totally different view of ancient Janssen from the one we had at Day 4, and its younger neighbors Fabricius and Metius.

*Day 19:* This is one of the finest nights to see all of Mare Tranquilitatis, the Sea of Tranquility, where the Apollo 11 astronauts landed. The Pyrenees Mountains dominate their part of the southern hemisphere. This is also the night for the old crater Fracastorius, as well as Piccolomini.

*Day 20:* Theophilus experiences sunset tonight, along with its neighbors Catharina and Cyrillus. These three craters offer a strikingly beautiful sight, and so do the Haemus Mountains and Menelaus. Mare Tranquilitatis is half gone now, and we can see on its surface Dionysius and Plinius, two well-defined craters. Can you find little Sabine? The Apollo 11 astronauts landed just east of there in July, 1969.

The tiny crater Linné is visible tonight, on the eastern part of Mare Serenitatis. This strange crater is more of a mound on which, in 1866, only a white spot was visible in the crater's position. The small craterlet became visible early in 1867, and has been visible ever since. Perhaps Linné is the site of some lunar activity, although evidence for that is uncertain.

*Day 21:* This is the night of mountains; both the Caucasus and Apennines are casting long shadows. More than any other lunar range, the Apennines seem to represent Earth mountains, except for the small craters you may find strewn around that range. The Mare Imbrium is also prominent this night, and we can guess that it may be the remains of a huge impact crater. Kepler, Copernicus and Tycho are marvellous. Although Archimedes is still lit by a relatively high Sun, its large neighbors Aristillus and Autolycus are beautifully placed for observation at sunset. Mare Imbrium is delightful tonight, and rising from its surface is a tall mountain west of crater Calippus that soars to over 17 000 feet. (Calipus added, around 340 BC, seven spheres to Eudoxus's scheme of 27 spheres for explaining planetary motion.) We also get a good view of Heraclitus and Licetus, a pair of craters showing interesting structure.

*Day 22:* Sunset over the Apennines, a range of mountains whose appearance changes markedly during the lunar month. At midmorning they seem flattened, near noon they show some dark spotty features, and now the range shows very sharply. We also get a good view of the Spitzbergen Mountains and Piton. On this night we can travel along the terminator of the last quarter Moon. The Sinus Medii and Rhaeticus define the equator. Rhaeticus is an example of a lava-filled crater, a little like Wargentin.

*Day 23:* This night offers two unforgettable sights. One is Mare Imbrium entering shadow. The second is Copernicus, a magnificent, complex sight at sunset. Also, we can say goodnight to Archimedes, and turn our attention to Mare Nubium, which is named appropriately 'Sea of Clouds' for all its ghost features. The walled plain Deslandres is right on the terminator. This is also the last night for crater Eratosthenes, as well as the last good view of Clavius.

*Day 24:* By now the Moon rises so late that you need some dedication to get up and see it. It is illuminating so little of the sky that people who search for comets and look at galaxies now can pay attention to the morning sky. Near tonight's terminator we can spot the Laplace Promontory, named for the French astronomer who developed the nebular hypothesis in 1796, at the junction of Sinus Iridium and Mare Imbrium. We get a last look at the Riphaeus Mountains, and huge Oceanus Procellarum begins to show some of its detail.

*Day 25:* This is the night of sunset on Aristarchus and the Cobra's Head. If you fight back sleep to see the waning crescent, you won't be disappointed. The earthshine is bright, so the entire surface can be seen. Sinus Iridium dominates the northern part of the crescent; it looks like half of some gigantic crater, and it is easily seen through binoculars. We still see Grimaldi toward the west, the Heraclides Promontory, and the sharp peaks of the Jura Mountains. Adjoining Grimaldi is the large walled plain Riccioli, a 100 mile long feature with dark patches.

Riccioli is a somewhat diminutive feature when compared to the reputation of the astronomer it honors. Joannes Baptista Riccioli was a Jesuit astronomer who in 1651 prepared a lunar map. Most of the names he gave to craters on that map are still used.

The southern limb is dominated by Gassendi, a beautiful sight this night. This crater is named for Pierre Gassendi, a famous mathematician and astronomer.

*Day 26:* Remember that last night Grimaldi portrayed its dark floor? With changing light near sunset, the floor is not nearly so dark. A contrasting pair of craters, dark-floored Billy and its bright-floored companion, Hansteen, offer a viewing challenge.

*Day 27:* You will be lucky to see the Moon at all tonight. This feat is possible mostly in late summer, when the thin crescent is highest and best placed. Although Grimaldi is barely visible, we can locate Hevelius to its north. This feature is named for Johannes Hevelius whose *Selenographia* of 1647 first demonstrated the effect of libration. We may even see Pythagoras dominating the north limb, although this depends somewhat on the effects of libration. Also on the north limb we can see a long oval walled plain called Otto Struve.

*Day 28:* With the finest sky, in late summer or early autumn, you may get a chance to see a Moon less than a day and a half from new. Your best instrument to try this feat is a pair of binoculars or a telescope with a wide field of view.

# 7    Moon II: Advanced observations

## 7.1    Crater drawing program

Once you have enjoyed the Moon and its many features as it orbits the Earth a few times, you might be interested to begin drawing. Your new purpose is to record each of the following seven features three times: once when the Sun rises over the feature, again at mid-day, and a third time as the Sun sets over it. Give yourself 10 or 15 minutes to examine the crater thoroughly and another 15 for your drawing. This project should give you a sense of the cosmic play of sunlight on an object in space. I suggest the following seven features:

(1)    Tycho – easily found, the center of the Moon's most dramatic ray system.
(2)    Copernicus – a huge, easily spotted feature with a fine central peak and sharp walls.
(3)    Plato – more of a 'walled plain' than a crater, this object is easy to draw because of its easily seen structure.
(4)    Archimedes – easily found on the eastern edge of Mare Imbrium.
(5)    Theophilus – a little more challenging and complex.
(6)    Alphonsus – harder to locate and quite complex to draw.
(7)    Apennine Mountain range – on the southeastern 'shore' of Mare Imbrium, a complex and beautiful range of mountains.

### 7.1.1    Drawing a feature

Lunar feature sketches are the best way to get started with astronomical drawing. The tones, colors, and shapes are usually very clear and distinguishable, and you can use your lunar sketching experience when you start to sketch planets, comets, and deep sky objects.

You do not have to be an artist to draw a lunar feature. In fact, it may help if you are not. What you need to acquire is not the interpretive skill of an artist but the ability to record data accurately and objectively. As an example,

I know of three photographs of Kitt Peak National Observatory. One is a truly marvellous picture showing the Sun setting over the McMath Solar Telescope. The other is of a few old trailers that make up Kitt Peak's maintenance area, and a tiny speck of light glimmering off a distant dome is the only indication you have of the presence of a major observing facility. In a sense, neither picture shows Kitt Peak for what it really is. The setting Sun is one of the worst times for observing seriously at Kitt Peak, and the little trailers are not where the observatory's action really happens. Neither photograph has captured the essence of the observatory. On the other hand, the third photo, a snapshot of an opened solar telescope taken by a child in the middle of a sunny afternoon, would more accurately reflect what really happens at this working institution.

Drawing a lunar feature requires the skill of the third photographer, who came up to the observatory and photographed what she saw. Your drawings will improve rapidly once you understand the kind of drawings you need to make. Simply record what you see, taking care to record the delicate interplay between sunlight and shadow on a lunar feature.

When should you do a drawing? Full Moon is absolutely the worst time for any feature that requires shadow contrast to be seen easily, although it is the best time to study tones and colors, and to sketch rays. Unless you are drawing one of the great ray systems, something near a limb, something which may be visible only around full Moon, this is a time just to watch and enjoy.

After the Moon leaves this full phase, shadows make the features more interesting. The setting Sun causes shadows to form, making a previously invisible feature suddenly appear, begging for your attention. Before you begin drawing, check your prospect. Observe it first with low power, then raise the magnification until the atmospheric seeing begins to degrade the image, making it fuzzy and its details hard to see. When you are using such a power, then go to your next lowest power eyepiece. The image you see should be clearer. The decision you have to make is to use magnification that lets you see details clearly, and this power will change from night to night, depending on how good the seeing is.

Are drawings better than photographs? Ideally, they should be, since most photographs are exposed over a period of time during which the currents of our atmosphere would flow by, obscuring faint details. Some nights this lack of steadiness, which we record as 'seeing', is more severe than on other nights. However, a drawing done by an observer is subject to the mental interpretation of reproducing on paper what is seen through the eye. It has been said that an observer can see details with a telescope one half or less the size of what can be recorded photographically; i.e. a 6-inch reflector can 'see' what a 9-inch can photograph, but an argument like that is overly simplistic. Experienced observers with carefully trained eyes can see much more than beginners, and a good telescope with good film on a night of good seeing can take a photograph that is hard to beat. In essence, the question

'Which is better?' should be replaced with 'Which would we rather do, drawings or photographs?', and we should proceed in the direction our gifts lie.

The first step is to be comfortable. Although I have seen a good observer make a good drawing through a refractor whose eyepiece was one foot from and facing the ground, and I had to help him rise from his supine position after he was done, I didn't see why he had to go through all that fuss when a tube that rotated in the saddle of its mount would have made him a lot more comfortable.

Once you have an eyepiece position that is at the right angle (facing you, not the ground), try to find a chair of the right height that you can use for still greater comfort. I have found that when comfortable, I can see stars up to half a magnitude fainter and subtle details on the Moon and planets that have eluded me before.

The next step is to pace your drawing. The old saying 'The stars wait for no one' is especially true for the Moon, whose surface features can change appearance within half an hour. Thus, for the first few minutes, just look at the crater and try to understand what is happening. Note the bright and dark portions and the relative positions of the most important features; the rim, the central peak, and any strange details on the floor.

Begin your drawing with a quick sketch of the positions of all these features. Outline the features carefully. It is the timing of this sketch, which marks where everything is, that is very important, so record the time you start and the time you finish the sketch. The midpoint of that is the time of your whole drawing.

Once you have completed the positional sketch, you can take your time filling in the details, using your memory supplemented by continued looks through the eyepiece. For a first drawing, take all the time you need. With experience you will probably find that ten minutes suffices for the sketch, and half an hour for a complete drawing, depending of course on how complex the area is that you are sketching.

Now that we have determined the time you should allow for a complete drawing, with what part of the lunar feature should you begin? This depends on whether your feature is on the sunrise or the sunset terminator. At sunrise, the trick is to draw the areas farthest from the terminator first, as their contrast is reducing rapidly due to the rising Sun. Because you are working toward the terminator, the exciting features at the edge will get drawn last. In this way, you automatically get to study these challenging parts a while longer as you draw the other areas.

For sunset drawings, you definitely need to draw the complex shadow areas at the terminator first, otherwise they will disappear into darkness before you get to them. Once you have drawn these, you can relax somewhat as you capture on paper the less contrasty areas further from the terminator.

While you draw, it helps to estimate quantitatively the relative brightness intensities of the features you draw. Try using an intensity scale such as this:

| 0 | the darkest shadows |
|---|---|
| 1 | dark grey areas near terminator |
| 2,3 | dark areas near sunset but in sunlight |
| 4,5 | normal crater floors near sunset |
| 6,7 | normal crater floor intensities under sunlight |
| 8 | bright ray systems under sunlight |
| 9 | mountains and mountain peaks in sunlight |
| 10 | the brightest spots, like Aristarchus |

While observers have tried many things, including charcoal, I have found that a medium soft (2B) pencil records details well. A good quality artist's stub will help to merge your pencil lines into a smooth shading, although soft tissues wrapped around the eraser end of a pencil sometimes work better. Although some observers ink over drawings after they are completed with a fine mapping ink, I do not think that is necessary.

I prefer drawing features without looking at any photograph or printed sketch beforehand. Otherwise I might draw what I think should be there rather than what I really see.

### 7.1.2   A note about notes

Help the people who might study your drawing later by recording everything you can about it. The midpoint time of your sketch, the type, size, and focal ratio of your telescope, the magnifications and the sizes of the fields of view are all important to record. You should note also as much as you can about the observing conditions. The seeing, a quantitative record of the steadiness of the sky, is a vital record because a night of poor seeing will cause you to miss much critical detail. The transparency, which records how clear the sky is, should also be noted, although in drawings of the bright lunar surface this value is not as important as seeing. (For a more complete discussion on seeing and transparency, see Chapter 9.)

## 7.2   Photographing the Moon

One August night I took a twin lens camera, focused it at infinity, mounted it on its tripod, and set it up so that its picture lens was virtually resting against the low power eyepiece of Pegasus, my 8-inch telescope. Using a slow color slide film, I took exposures with a wide open lens setting, ranging from 1 second to 1/500 second. I took the film to the drug store whose laboratory specialized more in beach pictures than in serious astrophotography. Since my camera was just resting against the eyepiece, not even attached to it, I did not expect anything significant.

The results were surprising. The 1/25 second exposure showed considerable detail along the terminator, with craters, a mountain range, and long shadows. My picture is reproduced in Figure 7.1.

Taking photographs of high quality is much more challenging. The most commonly acceptable camera for this work is a 35 mm single lens reflex with a removable lens. By attaching this camera with an adapter to the eyepiece drawtube of your telescope, the telescope itself becomes the camera lens. With a single lens reflex, you simply focus the Moon in the camera just as you would any Earth picture. Try several exposures at various speeds; with one of them you should get a good photograph.

It is very important to make sure that your telescope does not vibrate during the exposure. Even a tiny wobble will result in a fuzzy picture. In fact, when you press the shutter release button on your camera, you will probably impart so much motion to the telescope that your image will be blurred. A cable release will help. I suggest that you simply set the camera on its self-timer for 10 seconds or so; by the time the camera snaps the picture, the telescope will have settled down. The only problem with this solution is that unless your telescope has a motor drive (really a necessary part of the telescope even for lunar photography), the part of the Moon you photograph will not be the same as the part you targeted on earlier. Some planning will solve this riddle. Estimate the time you need between the moment you are satisfied with what appears in the eyepiece and the moment your camera

Figure 7.1. The Moon, taken through a 20 cm reflector with camera resting on the eyepiece. 1/25 s, Ektachrome X (ISO 64). Photo by the author.

shoots the picture after its 5 to 15 second wait. Then set the telescope an appropriate distance west of the Moon. By trial and error, eventually you will know exactly where to set the telescope so that the terminator will be in just the right place when the shutter is released.

### 7.2.1   At the prime focus

The rays of light a telescope's mirror reflects are focused at a point we call the prime focus. Since this position is directly in front of the mirror, putting your eye at that point would be the best solution, except for the problem that your head is so large that it will block most of the light trying to reach the mirror. Herschel solved this problem by having the mirror reflect rays at a slight angle. Newton's solution, to insert a small flat mirror a bit before the rays reach their focus, has lasted to the present day. The small mirror directs the rays conveniently to the side of the telescope, so that your head is safely out of the way of the incoming light path. Attaching a camera to the prime focus presents the same problem, so we attach it to the side of the telescope at the Newtonian focus. The focused rays strike the film directly, with no need for an eyepiece. The advantage of this approach is that you get a bright image that requires the shortest exposure time. If you want an image that is more highly magnified you can install an eyepiece into your system, so that the Moon's rays actually pass through an eyepiece before they reach the focus of your camera. The eyepiece projects its image of the Moon onto the film. You can see this effect for yourself by inserting an eyepiece and placing a piece of paper near the eyepiece so that an image of the Moon is projected against the paper. The farther away the paper is from the eyepiece, the larger the image. The camera film would 'see' the image the same way. Notice also that the farther away the paper, the fainter the image. With a camera, that means you need to expose the film longer, thus increasing chances for a blurry result.

Video cameras are quite useful for lunar photography, especially for monitoring specific areas. With an f/4 or faster system, it is possible to record the earthshine on the Moon's unlit portion, and search for anomolous brightenings known as transient phenomena.

## 7.3   Lunar transient phenomena

Is the Moon still a changing place? It may well have been on June 25, 1178, when five monks observed the remarkable sight of a young crescent Moon whose 'horns were tilted towards the east. Suddenly the horn split in two. From the midpoint of the division, a flaming torch sprang up, spewing out fire, hot coals, and sparks.' Observers must sometimes be patient before their work is published. This rather early possible report of a lunar transient phenomenon did not receive international acclaim until centuries later, when it appeared in Carl Sagan's *Cosmos*.[1]

1 C. Sagan, *Cosmos*, p. 85. New York: Random House, 1980.

Whether these monks had observed the formation of the young crater Bruno, as has been postulated, we cannot be certain. What is certain is that a careful watch for perhaps less impressive events is an important activity in which amateurs can participate. Activity on the Moon is very sporadic, and almost every one of your watches will produce nothing. However, the list of lunar transient phenomena dates back through several centuries. On November 26, 1668, there was a bright star-like point on the unlit side. During the lunar eclipse of December 10, 1685, Plato's floor showed a curious reddish streak, and during the total eclipse of August 16, 1725, a bright spot appeared on the lunar disk. The peak of Alphonsus apparently gave the astronomer Nikolai Kozyrev a fascinating night as he watched it show reddish, then whitish, hues. The spectra he took on the night of November 2, 1958, show evidence of outgassing on the lunar surface. Although some lunar scientists are skeptical, the possibility of this event being genuine was enough to choose that crater as the Ranger IX impact site.[2] More recent possible events were documented by both ground-based observers and the Apollo astronauts, including the entire Apollo 11 crew which saw an event in Aristarchus.

There are two ways of searching for such events. With virtually any size telescope, you can scan the entire surface of the Moon, bright and dark, in the hope of catching something bright. While such a procedure might be fun, it is not the most efficient. Get to know a specific region or crater, like Aristarchus, Plato, or Grimaldi, and concentrate on it. Unfortunately, a long and uninterrupted gaze could produce afterimages that could mimic transient phenomena; that is one reason why it is important that these events be observed by at least two people simultaneously.

Reliable observers have reported color changes, which may be as simple as the occasional appearance of colored hues, especially around the bright crater Aristarchus. A simple way to detect color phenomena involves a blink device that consists simply of a red and blue filter, mounted in a support such that you can easily move one filter, then the other, in front of your eyepiece.

Another fairly common report concerns a brightening or fading of a part of a crater in a process that is not obviously related to the changing shadow scene. Something else that could occur is an obscuration or hazing over of a region. Experienced observers have detected small regions that seem to get partially obscured while their surroundings remain sharp. Finally, there are the rare flashes. All these events, especially flashes, are spotted more easily on the dark side of a crescent Moon, the part lit by earthshine.

A severe problem exists in getting your observation of a transient event confirmed. With some luck you might have a second observer somewhere in the world who sees the same event. Astronomy clubs could set up common

2 R. L. Heacock, G. P. Kuiper, E. M. Shoemaker, H. C. Urey and E. A. Whitaker, *Technical report No. 32–800: Ranger VIII and IX*, p. 2. Pasadena: Jet Propulsion Laboratory, 1966.

times, especially during new or old crescent phase, when the Moon's unlit side is easily visible. If a large group of people are observing independently, chances for a confirmation are much better: make sure that the observations are independent.

There may be another way of getting confirmation, but although it is not as pure as having someone else observing from the start, it is legitimate. If you suspect something, call another observer with a request to observe the Moon. Say that you have seen something, but do not say what type of phenomenon and where it appeared.

The worst thing you can do is to submit an erroneous observation of an event. If you see a strange color on a crater floor, look at the floors of nearby craters that are subject to the same play of light. Is the strange color there also? If it is, then you are not looking at what is strictly a transient event.

### 7.3.1 Suspect areas

More than one observer has reported an LTP in the areas listed below.[3] These areas are plotted on Figure 6.1.

> Alphonsus
> Aristarchus, Herodotus, Cobra Head
> Aristoteles
> Calippus
> Cassini
> Clavius
> Cleomedes
> Copernicus
> Mare Crisium, near crater Peirce
> Eratosthenes
> Fracastorius
> Gassendi
> Sinus Iridium
> Kepler
> Lambert
> Linné
> Manilius
> Picard
> Pico
> Piton
> Plato
> Posidonius
> Proclus
> Riccioli

---

3  List adapted from John Westfall, *Lunar Photoelectric Photometry Handbook*. Association of Lunar and Planetary Observers, 1984.

 Schickard
 Taruntius
 Theophilus
 Timocharis
 Tycho
 Zagut

Despite large amounts of work on this problem of active lunar sites, the subject is still controversial. The Apollo spacecraft left seismographs liberally spaced across the Moon's surface, instruments that precisely recorded events like Saturn IV-B rockets crashing. They did record some moonquakes centered deep below the surface.

Some recent studies indicate that the most recent basaltic volcanism may have occurred as recently as one billion years ago. That seems like a long time, but it is four fifths of the amount of time that elapsed between the Moon's formation and the present. It is possible that the Moon still has a few active areas.

# 7.4 Notes on advanced projects

Since these pages are meant for beginners, the following notes on advanced projects are designed only to give you clues as to what you can get into as you become more experienced. Contact an astronomical friend, society, or the Association of Lunar and Planetary Observers (ALPO; see Chapter 24) for more information.

## 7.4.1 Lunar height measurements

Another lunar program for trained observers involves the use of trigonometric functions and careful photographs or drawings of the shadows of mountain peaks to obtain height measurements of lunar mountains. While the heights of lunar mountains are well known, this project is an interesting exercise that enables you, with your own telescope from your own back yard, to determine the height of a mountain on another world.

The procedure is actually easier than a look at the formulae would indicate. It does involve some trigonometry. The idea is to estimate the length of the shadow a feature makes on a nearby flat area by comparing it to the unforshortened diameter of a nearby crater. If the 'flat area' is slightly tilted or inclined, the length of the shadow will be longer and will fool you into thinking that the mountain is higher than it really is.

First we need these data:

 the time of the observation to the nearest minute;
 the location of the object to be measured;
 the identity of the comparison crater.

Then we calculate the measurement using the following formula:

$$H = fD \sin A \operatorname{cosec} F - \tfrac{1}{2} f^2 D^2 \operatorname{cosec}^2 F \cos^2 A,$$

where

> $f$ = your estimate of the shadow's length, expressed as a fraction of the comparison crater's diameter;
> $D$ = diameter of the comparison crater, expressed as a fraction of the Moon's radius;
> $A$ = the angle of solar altitude as seen from the mountain;
> $F$ = angle between Earth and Sun as seen from the Moon's center.

How do we calculate $A$?

$$\sin A = \sin B \sin b + \cos B \cos b \sin (L - C),$$

where

> $B$ = mountain's selenographic latitude;
> $L$ = mountain's selenographic longitude;
> $b$ = Sun's selenographic latitude;
> $C$ = Sun's selenographic colongitude. (A way of indicating the position of the sunrise terminator on the Moon. It is equal to the sunrise terminator's selenographic longitude, measured eastward from the center of the Moon's disk. It increases at a rate of about $\tfrac{1}{2}$ degree per hour.)

How do we find $F$?

$$\cos F = \sin B' \sin b + \cos B' \cos b \sin (L' - C),$$

where

> $B'$ = Earth's selenographic latitude;
> $L'$ = Earth's selenographic longitude.

We can find $C$, $b$, $L'$ and $B'$ from the *ALPO Solar System Ephemeris* published each year by the Association of Lunar and Planetary Observers, or the *Astronomical Almanac*. When we are finished, we have the height $H$ in terms of the Moon's radius, and by multiplying by 1 738 000 we can get the result in meters.

We can calculate the angle of slope of a lunar feature very easily by recording the time at which the shadow disappears. The solar altitude $A$ is the same as the angle of the feature's slope.

### 7.4.2    Viewing difficult features

Occasionally subtle features become visible for brief periods when the lighting angle is proper. For example, when the Sun's selenographic colongitude is between 14.10 degrees and 14.40, some observers have reported a low dome on the floor of Plato. This is not a transient phenomenon, just a regular feature visible only at certain angles of sunlight.

You probably think that with the Surveyors, Lunar Orbiters, and Apollo missions, the Moon must surely by now be completely charted. It is not. Small sections at the extreme south limb are very difficult to see from Earth and have not been charted even by the Lunar Orbiters. This mysterious area is known as 'Luna Incognita' and it has been a project of the ALPO Lunar Section to map it.

Luna Incognita is a large area on the Moon's south and southwest limbs. It is south of the craters Short, Newton, and Drygalski, and on the southwest limb near the large crater Bailly and crater Hausen. These areas are visible when the Moon's libration is favorable, with a strong southerly libration preferably combined with a strong westerly one.

Since Luna Incognita is so poorly observed, I suggest that you always begin a drawing by noting the details of the nearby and better known features like Bailly. When you are confident of the positions of these features, then you are ready to set off into the uncharted region of Luna Incognita. It is very important to plot the positions of features or suspected features as accurately as possible.

**Photoelectric photometry** Photometry of the Moon? Most observers think of photometry as a tool for measuring the brightnesses of faint objects, like minor planets and variable stars. Just think, however, of the chance you might have to record, quantitatively, the changing brightness of a lunar transient phenomenon like a brightness change in Alphonsus. Also, just as we can learn about the nature of a minor planet or a star through measuring its light, we can also measure the light on different parts of the Moon to learn more about its physical composition.

Measuring the Moon's ray systems is a particularly useful function for photometry, since the rays are some of the youngest features on the Moon's surface and they seem to change brightness in a different pattern from what appears as usual elsewhere on the satellite. In the Moon's case, 'usual' is defined as what is typical for nearby sites having the same brightness relative to each other throughout a lunar month.

Color is another useful area that a photometrist can study. Photometry through different filters can measure color differences, and thus lead to a better understanding of the composition of different lunar surfaces.

What kind of lunar sites are best for photometry? First, you don't want a mountainous or cratered site with many shadows. You need a place that can be readily identified under varying amounts of sunlight, so that you can center your photometer consistently. Also, the aperture of your photometer should be set for about 10 km at the Moon's distance. This is the equivalent of a very high power eyepiece, but the Moon is so bright that the smaller the area, the less scattered light you will have to worry about. Also to reduce scattered light, your telescope's optics need to be very clean, and the sky transparency should be excellent, with no clouds or haze. If what you are measuring is near the terminator, the lower overall light level makes the effect of scattered light more severe. Also, the smaller the area you measure, the more severe the problem with scattered light.

# 8   The Sun

**WARNING!**
**Observing the Sun is dangerous. Unless you have proper protection for your eyes, permanent blindness can result from the shortest look through binoculars or a telescope. The eyepiece filters that often are supplied with small telescopes are not safe.**

Near the western border of Cetus the Whale is a faint star known unpretentiously as 9 Ceti. Barely visible to the unaided eye on a dark night between Beta and Iota Ceti, 9 Ceti is most easily observed during the evenings of August and September. Further east, in the northern part of Orion, is a brighter star called Chi 1 Orionis. Neither star dominates the sky, but they share something rather important. They are very much like the Sun.

If we were to trade places and travel to either star, an Earth-like planet nearby would have to provide a thicker protective coat of ozone than ours, for the middle layers of these stars, their chromospheres, are more active than that of our Sun. In fact, it is difficult to find an exact solar clone.

What provides all the heat, light, and other energy to support life on our planet is a simple, single G2 dwarf called the Sun. We can actually observe several G stars; although these stars tend to be undistinguished beacons in the sky, far in intensity from the brightest stars, Capella is a pair of giant G8 stars. As night ends, the G2 that rises is well worth observing and studying closely. Whether we worship it, plan our lives by its schedule, tan ourselves by its light, bask in its warmth, or study it, this G2 star called Sun has an importance that cannot be overstated.

Close and always here, the Sun is convenient to study. It is so easy to find that we take it for granted, spending our nights observing the stars of great distance, while during the day this brightest star of all shines ignored by most astronomers.

My grandfather and I first observed the Sun on the roof of his apartment building at the end of a hot afternoon during the summer of 1960. I remember the look of sheer amazement on his face as we noted the two long strings of sunspots that lined both hemispheres of the Sun. How lucky we were to meet the Sun on such a day, for in 1963, when I began my first serious spot counting program, I was in for a different story. Just one year from sunspot minimum, there was but one tiny spot on the entire surface! In the months after that the Sun often went through long periods of time showing no sunspot activity at all.

By the spring of 1966, solar activity was in a renaissance. With the level of activity increasing by the week, it was easy to choose a day with many spots. In Montreal, interest was not for the Sun but for the construction of a World's Fair set to open a year later. One morning I walked to a site that had coin-operated binoculars so that people could inspect the fair site.

Just as I arrived, the sky cleared briefly. Worried that it would cloud over again before I got home to record my daily view of the Sun, I quickly inserted a dime into the binoculars and pointed them upward, projecting the Sun's image on a bus transfer which was the only paper I had with me. Sure enough, the binoculars showed two spots. I cannot remember seeing any World's Fair buildings, but my memory of the sunspot observation means that the dime was well spent!

## 8.1   Observing the Sun is dangerous

The key to pleasant solar study is to do it in safety. Do you own a small refractor with a Sun filter that attaches to its eyepiece? You do? Then, put this book down, find a hammer, and destroy the filter! Eyepiece filters are dangerous because they work where the Sun's focused rays are hottest. All that concentration of light is likely to break the filter sooner or later, probably at the moment your eye is next to it. The resulting eyeful of Sun will damage your eye permanently in an instant.

When I am observing the Sun with a group of children, I always begin by demonstrating how dangerous looking at the Sun really is. Producing a cheap plastic garbage bag (they, and the human eye, are damaged faster than the expensive ones) I ask the children to imagine that their eyes are the equivalent of the bag. I then move the bag to the focus point of the eyepiece, the place where the Sun's rays are hottest, and the place where the observing eye is placed, and ask the children to count the seconds until the bag ignites.

Especially if the sky is not hazy or covered with light cloud, the plastic should ignite almost instantly, burning a clean hole. I then let the Sun carve its initials into the material, and offer it as a souvenir. When the experiment is over, I then ask if anyone still wants to look directly into the unprotected eyepiece. 'Are you crazy?' the children inquire, and I know that their lesson is learned.

How do you safely find the sun? First, do not use the finder. Make sure, in fact, that the telescope's finder is either covered or removed so that no one accidentally looks through it. Find the Sun by moving the telescope until its shadow is as small as possible. The Sun should then be close to shining its rays right down the tube.

There are only two safe ways to look at the Sun: projection, and direct viewing with a filter that covers your telescope's *front end*. With projection, you point your telescope towards the Sun and allow the rays to project through the eyepiece onto a piece of white cardboard or white paper at the bottom of a cardboard box. In this way you can achieve an image of the Sun over 60 cm in diameter which several people can observe at once. The advantage for teachers is obvious. Children can see the projected image all at the same time, and the whole process need not take more than five minutes of their time. As well, it can be repeated day after day as the spots march across the Sun's surface.

The other safe way is to use a filter placed not at the eyepiece but at the front end of the telescope, blocking off much of the Sun's light before it even begins to be focused. With this approach, remember that the filter must be placed at the front end of the telescope tube, before the Sun's rays hit the mirror or objective lens and see any magnification. The advantage of the filter method is that the telescope's optics do not get at all hot, and if the filter is of good quality it will deliver an image of the Sun that you can view safely and comfortably through your eyepiece.

Make certain that the filter material is of good quality. The major astronomical magazines carry advertisements by several companies that sell filters, some of glass, others of an aluminized mylar material. Although I have tested some and find the mylar to be comfortable, the material yielding a blue Sun image, the glass ones should be safe too.

No matter what filter you choose, be very careful. If the material is torn, or the coating on the glass is scratched, do not use it. If you look through a filter that has worked well and suddenly lets in too much light, discard it. Once again: the only safe filters are those designed to fit at the front end of your telescope. Those that fit at the eyepiece of an otherwise unprotected telescope are not safe.

Otherwise unprotected? There is a third method, the Herschel Wedge, that until the early 1970s was rather popular. However, recent tests indicate that even with their accompanied filters, these wedges let in too much ultraviolet and infrared to be considered very safe. This approach involved actually replacing the secondary mirror of a Newtonian reflector, or the zenith prism on a refractor, with a wedge-shaped glass that harmlessly transmits almost all the Sun's light. What is reflected back, less than 5% of the Sun's energy, goes through an eyepiece filter. Since only this small percentage is transmitted past the wedge, the use of an eyepiece filter in this restricted manner is supposedly safe; with so much less light, the filter is not likely to crack, and even if it does you have a fraction of a second to protect your eyes. With the coming of good front-end filters, Herschel Wedges are out of fashion these days.

## 8.2    Observing projects

### 8.2.1    Daily sunspot count

The most obvious solar features are the *sunspots*, magnetic storms on the solar surface that appear dark because they are cooler than the rest of the surface. These spots usually appear in groups, and if you look carefully around these groups, you may notice some brighter regions called *faculae*, especially near the Sun's limb or edge.

Who really discovered the sunspots? Although the Chinese recorded naked-eye dark features in 28 BC, and other observers, including Kepler, suspected that these events were transits of Mercury or Venus, Galileo observed them systematically for several weeks before concluding that they had to be events taking place on the solar surface. Eventually he wrote this controversial letter:

> Repeated observations have finally convinced me that these spots are substances on the surface of the solar body where they are continuously produced and where they are also dissolved, some in shorter and others in longer periods. And by the rotation of the Sun, which completes its period in about a lunar month, they are carried around the Sun, an important occurrence in itself and still more so for its significance.

The concept of discovery requires that the discoverer understand the significance of what has been seen. Thus, we can credit Galileo with the discovery of the sunspots.

Even Galileo recognized that the sunspot numbers varied from day to day, but observers watched the Sun for over 200 years before the next really significant discovery occurred. In Dessau, Germany, Heinrich Schwabe in 1826 bought a small refractor telescope for himself. Not being sure what useful work he could do, he sought the advice of an astronomer who suggested the Sun. Perhaps he would spot the elusive Vulcan, a hypothetical planet whose secret would be given away when it crossed the disk of the Sun.

The project was not terribly time consuming, as Schwabe could undertake it at his convenience during the slow afternoon hours. While waiting for Vulcan to appear he took meticulous records of the numbers of spots. Seventeen years later, in 1843, he announced a possible sunspot cycle lasting about a decade. The Gold Medal of the Royal Astronomical Society went to amateur Schwabe in 1857, the man who had begun a project uncertain as to how it would end. Vulcan was never found, but the cycle of sunspot activity is a highly important find.

We have some questions about the cycle even today. We know that the period averages about 11 years, but the time between one maximum and the next can range from 7.3 to 16 years, and from one minimum to the next can range from 9 to 13.6 years. This means that the time between a minimum and a maximum to another minimum can vary widely. In fact, the traditional behavior involves a rapid rise over two or three years from minimum to maximum, followed by a much more leisurely decline.

Not long after Schwabe's find, Christopher Carrington made another important discovery. As a new cycle begins, the few spots that form tend to occupy high latitudes, and as the cycle progresses towards maximum and intensifies, the spots appear at more 'temperate' latitudes on both solar hemispheres. A cycle ends with a few spots close to the equator. Sometimes when we see some spots at the equator and a spot near the pole, we know that an old cycle is ending and a new one beginning and to which cycle the spots belong!

How do we study the sunspots? The easiest way is to count them, a procedure that is really as easy as it seems. To measure solar activity we need to recognize two things: that the Sun may have several disturbed areas, and that each of these areas may contain several spots. It is a bit like calling the weather service to inquire of the number of thunderstorms on the planet at this moment. While the number of individual thunderstorms is important, many of these storms are part of larger fronts or convective complexes, and that number of major systems is more significant.

Recognizing this, R. Wolf of Zurich, Switzerland, devised a method for recording the relative sunspot number for each day, a formula whose result would reflect not just individual sunspots but the wider picture of solar activity. It is a formula you can apply easily to your own observations:

$$R = k(10f + g),$$

where $R$ is the relative sunspot number, $f$ is the number of disturbed areas, defined as a group of spots or a lone spot unrelated to other spots, and $g$ is the number of spots, whether they be alone or part of groups. Thus, if you see only one spot, you have but one group ($10 \times 1$) with one spot ($+1$) and your $R$ number for the day is 11. On a more complex day, you might see four groups of spots, and a total of 57 spots. Thus, your $R$ number would be $4 \times 10 = 40$, $+57$, or $R = 97$. Each day, record the number of groups and the number of spots in all of the groups.

The '$k$' refers to a factor that considers the quality of the observer and the telescope being used. If you submit your observations, the group to which you send them will assign you a '$k$' factor. Meantime, give yourself a $k = 1$.

The '$k$' factor seems somewhat unscientific, but is needed since the $R$ sunspot number is obtained using the values of many observers. An observer with a 4-inch refractor will obviously see more than a person with a reflector half that size. However, it is in the averaging of many good observers, doing the best they can, that a valuable daily relative sunspot number is derived. Figure 8.1, the daily report form, asks for the '$R$' numbers for the Sun each day. There are probably more accurate ways of accomplishing this, but the formula has been used since the earliest days of sunspot counting and the system survives partly for the sake of consistency with older data. After all, a group of two spots covering a large area would get a number of 12, and if there are two tiny, separated spots on other parts of the Sun, your number leaps to 34, an increase of almost three times, due to two tiny spots!

In Greenwich, a somewhat different kind of record is kept which measures the area covered by all the spot groups. When these new data reductions are compared with the $R$ numbers, the result shows that the $R$

Figure 8.1. Report form for daily solar observations.

MONTREAL CENTRE
ROYAL ASTRONOMICAL SOCIETY OF CANADA
Solar Section: Summary Report Form

observer _____  month _____ 19 _____
address _____

| DAY | C SEE | D TRAN | E U.T. | F GR | G SP | R | J N-GR | K S-GR | L N-SP | M S-SP | REMARKS |
|---|---|---|---|---|---|---|---|---|---|---|---|
| 1 | | | | | | | | | | | |
| 2 | | | | | | | | | | | |
| 3 | | | | | | | | | | | |
| 4 | | | | | | | | | | | |
| 5 | | | | | | | | | | | |
| 6 | | | | | | | | | | | |
| 7 | | | | | | | | | | | |
| 8 | | | | | | | | | | | |
| 9 | | | | | | | | | | | |
| 10 | | | | | | | | | | | |
| 11 | | | | | | | | | | | |
| 12 | | | | | | | | | | | |
| 13 | | | | | | | | | | | |
| 14 | | | | | | | | | | | |
| 15 | | | | | | | | | | | |
| 16 | | | | | | | | | | | |
| 17 | | | | | | | | | | | |
| 18 | | | | | | | | | | | |
| 19 | | | | | | | | | | | |
| 20 | | | | | | | | | | | |
| 21 | | | | | | | | | | | |
| 22 | | | | | | | | | | | |
| 23 | | | | | | | | | | | |
| 24 | | | | | | | | | | | |
| 25 | | | | | | | | | | | |
| 26 | | | | | | | | | | | |
| 27 | | | | | | | | | | | |
| 28 | | | | | | | | | | | |
| 29 | | | | | | | | | | | |
| 30 | | | | | | | | | | | |
| 31 | | | | | | | | | | | |
| TOTAL | | | | | | | | | | | |

C – seeing
D – transparency
F – no. of groups

G – no. of spots
J – no. of groups N hemisphere
K – no. of groups S hemisphere

L – no. of spots N hemisphere
M – no. of spots S hemisphere
R – totals 10F plus G

number places too much emphasis on small changes in spot numbers. However, in their monthly or annual means, the number does present a good picture of solar activity and compares well with the Greenwich record.

Exactly what constitutes a sunspot group? Although in most cases it is not difficult to determine where one group ends and another begins, when groups become large and complex, the answer is not straightforward. A sunspot group is normally characterized by at least two large spots, known as a 'preceding' and a 'following'. These two spots are often accompanied by smaller ones.

A group evolves from the smallest detectable solar feature, a granule, which, as it begins to darken, becomes a pore. If a pore continues to darken, and expands, it becomes what is known as an 'umbral' sunspot. With further growth a lighter penumbra will form around it. Often several nearby pores darken concurrently, in an elliptical cluster. As the pores form sunspots they often divide into two regions, the 'preceding' one forming faster than the 'following' one.

At the height of its development, a sunspot group, with its two major sets of giant spots, can be quite a sight. As the group begins to fade, the following spot dissolves first, while the preceding one becomes round, loses its penumbra, and eventually disappears as well.

## 8.3   Other features on the Sun

When you get a good look at the active Sun you may be treated to much more than simple sunspots. On a very clear day with steady atmosphere and quality telescope you may see the Sun take on a mottled appearance. This is called *granulation*; the Sun's surface is covered with these convective cells that are separated from each other by slightly darker intergranular material. These cells are always being replaced with others, as each individual *granule* rarely lives longer than 10 or 15 minutes. When a granule is as dark as its surrounding material, it becomes a pore, a possible precursor to a sunspot.

On the Sun's limb, and around a sunspot group, look for obviously brighter patches or spots. These are called *faculae*. These bright areas often exist before a sunspot group is formed, and they outlast a group as well. Sometimes faculae may actually cross the surface of a spot, becoming a light bridge.

Occasionally a very bright spot appears on the Sun's surface near a major spot group. Those who have seen these *flares* assure the rest of us that you cannot miss such an event if it is taking place when you look at the Sun through your telescope. As seen in ordinary white light, these eruptions are very rare, reported once or twice each sunspot cycle, and do not last more than a few minutes.

### 8.3.1    Disk drawings

The ultimate sun tan is contained in a drawing of the Sun's disk. Imagine yourself on vacation at a summer campsite, where everyone is active in some sort of outdoor activity under a bright sunny sky. While your friends take advantage of the Sun, so do you, by drawing the placement, shape, and structure of the features you can see on the Sun. Chances are that your drawing might attract some attention, so you will have the chance to share your enthusiasm about the Sun as well as completing a useful drawing. As other people gather around, you become a teacher, showing off the Sun, and explaining that since even the smallest spots are a large fraction of the size of Earth, a large group marching across the Sun's face would cover an area several times that of our planet.

This chapter has a drawing form (Figure 8.2) that is designed for whole disk drawings of the Sun, and on it you draw the groups and spots as you see them through your protected telescope. Begin by studying the Sun's image. How many groups are there? Which is the most imposing? Are any small and hard to detect spots on the limb or near the center? When you think you have a good idea of the whole picture, then sketch the positions of each group on your paper. Since large, active groups may change shape slightly over as

Figure 8.2. Report form for whole disk drawing. The numbers represent where solar latitudes intersect the Sun's limb. Courtesy Solar Section of the Association of Lunar and Planetary Observers, noting that the ideal size of the circle should be 18 cm.

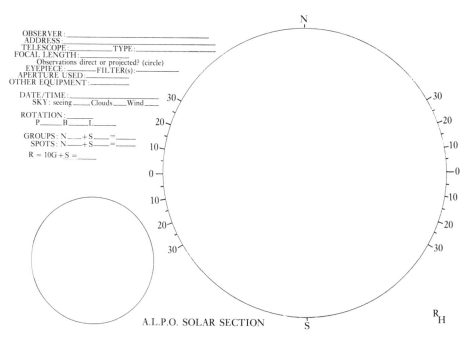

small a time as 30 minutes, you need to do this part as quickly as possible, using short dashes to mark the position. Even though you may spend a lot of time filling in the details, the time you put on your completed work is the time of the quick placement sketch. Be careful not to draw the spots inordinately large; it is unlikely that your giant spot group actually covers a third of the diameter of the Sun's disk! The outer, penumbral layers of sunspots are best recorded by marking their outer borders with a soft line and leaving their interiors unshaded. Outline the positions and shapes of faculae as well.

Your sketch completed, you then fill in the details, paying close attention to the degrees of darkness, the placement of small spots relative to the large groups, and the interaction of the bright faculae with the nearby sunspots. Finally, check carefully over the entire surface of the solar disk, again for tiny, hard-to-see spots that are far from the larger spots.

Next you find out the directions of east and west. This is easy. If your telescope has an equatorial mount, move it back and forth slightly on its right ascension axis (see Chapter 4) and watch one of the sunspots cross the moving field of view. Its direction of movement is the east–west line. Or, let the Earth's rotation do the work for you. Choose a sunspot and place it near one end of your field of view. Its motion across the field will then give away the east–west line, and you will know how the sunspots are oriented.

Figure 8.3. Report form completed by Rik Hill, Recorder of the ALPO Solar Section. The notation at lower left refers to types of sunspot groups.

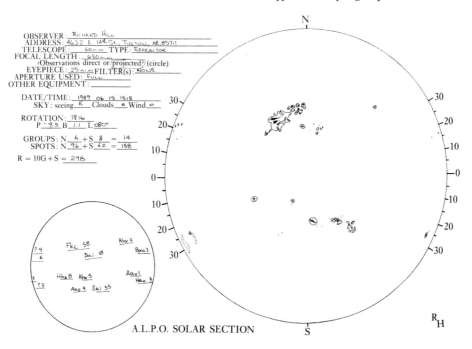

## 8.3.2   Detailed drawings

You can also draw a single group or a large spot in much greater detail by using Figure 8.4, the detail drawing form, and coding the spot you choose to draw to its corresponding appearance on the full disk drawing. Shade the spot umbrae rather thickly, and now you can shade lightly the penumbrae as well, but pay close attention to relative darknesses of these two parts of each sunspot. Outline the positions of any faculae that are associated with the spot group, and in the remarks note whether it is faint, bright, or intense.

How large is the feature you have drawn relative to the entire solar disk? The Earth's rotation can again provide you with an answer. Place the telescope such that an edge of the spot group is also at the edge of your field of view. (If your telescope has a clock drive, turn it off.) Depending on what edge you have chosen, the group will either leave the field or enter it; count the number of seconds until the other edge of the group crosses the edge of the field. Then count the number of seconds for the entire solar disk to cross the edge of the field. The ratio between the two times is the relative size of the spot group to the solar disk. Also, determine the east–west line just as you did with the disk drawing.

The aim of drawing is not to make you into an artist; instead, its purpose is to record data as accurately as possible, and, through such careful attention, to increase your understanding of the Sun's changing face.

## 8.3.3   Photographs

Photographs of either the whole disk of the Sun, or of the details of individual groups, are a good way to record a journal of solar activity. The procedures are simple. I have taken passable photographs of the projected image, using no filters at all. In any case, photography absolutely demands the most stable of telescope mountings. Even a 1/500 s exposure can result in a blurry image if the telescope is moving around because of a wobbly mount.

Figure 8.4. Detail drawing form: reproduce this on the top of a sheet of paper.

MONTREAL CENTRE
ROYAL ASTRONOMICAL SOCIETY OF CANADA
Solar Section: Detail Drawing Form

observer _____    date _____ 19_____
address _____    time: local _____ U.T._____
_____    seeing _____ transparency _____

reflector ☐   refractor ☐   direct ☐   projection ☐   aperture _____ magnification _____
fl objective _____    fl eyepiece _____    projection distance _____
relative diameter of solar disk _____    identification _____

remarks _____

Most observers doing serious work use a front end filter. Use a fine grain film like Kodak 2415 Technical Pan Film at about 100 ISO. Try for the shortest exposure possible; with the Sun there should be no shortage of light! At first you will need to experiment with your combination of telescope, front end filter, and film until you get the 'perfect' exposure on a day with no haze or light cloud covering the Sun.

Since they allow you to see the image before you photograph it, single lens reflex cameras are best for this work. If you can manually get the camera's mirror out of the way before you take the exposure, you will avoid blurring the image as a result of camera vibration.

Even when you think you have figured out the ideal exposure, take two or four more photographs using shorter and longer exposure times. These different exposures will show different types of detail; since the limb is fainter than the center, its features will show up better on longer exposures. Print your solar images on 8×10-inch paper, so that you will be able to study detail. Put north up and east to the left.

Taking detailed photographs is helpful to our understanding of daily changes on the Sun's photosphere. The best pictures are taken during moments of good seeing when individual granules are visible. If the seeing around professional sites is not sufficiently good on a certain day, then your photograph, submitted to the ALPO Solar Section (see Chapter 24), may provide the best evidence of what happened on that particular day.

## 8.4   Advanced work: hydrogen-alpha filters

Besides the spots, would you like to see the granulation much more clearly, along with filamentary structure? Would you like to see some *prominences*, tongues of gas which straddle the limbs on an active day, and a greater chance of flares than is possible in white light? This is what the Sun might show you in hydrogen-alpha light. Without H-alpha the only times we conveniently see these prominences are with a 'coronagraph' or during a total solar eclipse. At some expense, you can buy an H-alpha filter that will transform your white light Sun, with its sunspots, into a seething, churning furnace. You will probably want to study a big prominence several times during a day, to see it change its appearance as its gases leap off the Sun's surface and then return to it.

The Sun is a busy place, simple to observe, and its changing face is fascinating to watch. With a little practice you will soon become an addict as you study and enjoy what our star has to offer.

# 9   Jupiter

Late one evening during the summer of 1964, I was attempting to observe Jupiter through a 20 cm reflector. The giant planet was rising, and it had just cleared my neighbor's house – not an appropriate viewing time, since hot air rising from the roof would make Jupiter's appearance unsharp. This would be a quick look before bed.

In any event, the shimmering planet caught my attention more than I had expected, for it was a minute or so before I noticed a police car parked in front of the house. Two officers emerged and started walking toward me. From their almost military gait, I assumed that this would be an official visit. They quickly reached the telescope, and then halted. I looked at them; they looked at me.

One officer then broke the silence; Excuse me, sir, would you mind if a couple of nosy policemen looked at Jupiter?'

The brief look those men had that night showed an object that would have astonished ancient observers, and confirmed their view that it was king of planets. Its symbol represents a modified Z standing for Zeus. Jupiter leads our discussion of planets because it *usually* is the easiest to find and the richest to observe. The other planets are arranged here in the order of how easy each one is to find and begin to observe, easiest to hardest.

## 9.1   Jupiter and its moons

When Galileo first noticed the movements of three, and then four, objects near Jupiter, he realized that they had to be moons that orbit Jupiter in much the same way our own Moon orbits us. He was thrilled by these delicate movements and announced them enthusiastically. It was years later that these and other discoveries led the Roman Catholic Church to force him to recant, to deny the discoveries and their implications. The Earth must remain safely at the center of things.

The names of Galileo's moons are Io, Europa, Ganymede, and Callisto, and with much fainter Amalthea, found by Barnard in 1892, these are the only moons to have been discovered visually. The Voyager spacecrafts found exciting worlds – active volcanoes on Io, a smooth covering of ice on Europa, the craters and complex grooves of Ganymede, and heavily cratered Callisto.

Why not recall Galileo's work by recording the positions of the Galilean moons for a month or so? Such a project has no scientific value, of course, and you can even check your own identifications with the charts in *Sky and Telescope* or the Royal Astronomical Society of Canada's *Observer's Handbook*. But just this once, forget these printed charts and try to figure out which moon is which. Io will appear to move the most quickly, completing one orbit in just 1.77 days. After you have finished your 30-day moonwatch you can check one of the sources to see how well you did. If you have a primarily mathematical interest in observing, this observing project will acquaint you with the subject of orbital mechanics. If your interest is at all romantic, you will have just made four lifelong friends.

## 9.2    Seeing

Learning to see detail on a distant planet is really an art form, as William Herschel wrote over 200 years ago:

> Seeing is in some respects an art which must be learnt. To make a person see with such a power is nearly the same as if I were asked to make him play one of Handel's fugues upon the organ. Many a night have I been practicing to see, and it would be strange if one did not acquire a certain dexterity by such constant practice.[4]

To see any real detail even on this largest of planets, you need at least a good 10 cm refractor or a 15 cm reflector telescope. Smaller telescopes will show some detail, but not really enough to record. Remember also that good planetary observation requires that Earth's atmosphere be steady. Observing the details of planets requires a sharp eye that can pick up details at the very limit of visibility, like reading the words on distant road signs.

*Seeing* is a measure of the steadiness of the image of an object in the sky. If our atmosphere is unsteady, it will be impossible to detect these hard-to-see details. It is related to, though not the same as, *scintillation*, the rapid brightness changes we see in the twinkling of stars. Sometimes poor seeing results from turbulence in the upper atmosphere, and on other nights the problem may lie in the atmosphere just above you. On one night I was observing from a site high in some mountains. Although it was very windy, the seeing was good and planetary details were sharp and clear. Then I returned home and started observing again from my own site. The wind was gone and the session was much more comfortable, but the seeing had completely deteriorated! Probably I was trying to observe through that wind raging not far above me.

I have found that a hazy night usually is a still night with good 'seeing' for planets. Does this mean that the murky skies over cities on humid nights may be ideal for good planetary observation? Quite possibly; if the murk is swamping everything fainter than the planets and brightest stars, and if there are no strong upper-atmosphere winds, you might take advantage of a fine night for planetary observation.

Observers in Europe favor a scale developed by the planetary observer Eugenios Antoniadi (1870–1944), who devised a five-point system where 'I' represents a perfectly steady image, 'II' involves excellent moments lasting for several seconds, 'III' refers to average seeing where a good image is frequently interrupted by fuzzy periods, 'IV' involves almost constant 'fuzz-

4 W. Hoyt, *Planets X and Pluto*, p. 12. Tucson: University of Arizona Press, 1980.

ing out' of the image, and 'V' is so bad that planetary detail is not really visible at all.

There are several variations of a seeing scale that is measured in numbers from 0 to 10, where 0 represents a wholly unsatisfactory sky, so bad that Saturn's rings cannot be separated from the planet's disk; 5 represents some periods of steady viewing; and 9, rarely achieved in practice, represents long periods of totally steady images. Notice that the two scales are inconveniently in reverse order: if you use the first one; record the seeing in Roman numerals.

Admittedly these schemes are subjective. A steady night for an observer with a 7 cm telescope may be decidely unsteady for the same observer using a 30 cm telescope. However, it is a good basic guide to the atmosphere's steadiness, and most visual observers are satisfied with using it. In Chapter 15, on double stars, a more accurate seeing scale that uses the telescope's ability to separate double stars will be described.

## 9.3    The face of Jupiter

Things happen fast on Jupiter. The most challenging aspect of your Jupiter program comes when you are ready and willing to draw the disk of the giant planet. It is a challenge; one full Jovian Day is less than 10 hours long, and the planet is many times larger than Earth. Jupiter rotates so fast that it appears elongated, or oblate, even through a small telescope. Doing a drawing under such conditions requires strong concentration, as details will move noticeably across the planet's face in as short a period as 10 minutes.

Jupiter's face is consequently the most changing of any major planet. It consists of dark brownish strips called *belts*, interspersed by brighter *zones*. The belts are busy with smaller temporary features like *bridges* that seem to join one belt to another, and *festoons* which jut out from a belt.

How many of the dark belts and bright zones can you see? At first, all you are likely to notice are the two major dark regions known as the north and south equatorial belts. With a few minutes of concentration, other belts will slowly show themselves. Then try to see detail around and within the belts. Since the equatorial belts are the most active, you should be able first to see some of the festoons and other features that protrude from these belts. Sometimes a festoon will stretch almost all the way to another belt. If the seeing is especially good you may then notice some activity inside a belt. Look first for simple breaks inside each belt, then for streaks, spots, and other markings. The whole process of just looking at the planet should last around 10 or 15 minutes.

The mechanisms that power this marvellous atmosphere are very complex. Could the zones and belts, respectively, be large areas of updrafts and downdrafts in Jupiter's turbulent atmosphere? Until the Galileo probe studies Jupiter's vast interior, we do not know even whether these features are driven by the Sun, like winds on Earth, or by Jupiter's own considerable internal heat. When Jupiter was formed out of a hydrogen cloud, had its mass been only ten times greater than it was, it likely would have become a red dwarf star!

Those two policemen from the incident mentioned at the start of this chapter may well have observed a prominent reddish-brown spot that dominated the southern edge of the south equatorial belt, visible near the top of the planet, for northern hemisphere observers, in the inverted telescope field. The feature is called the *Great Red Spot*, and its huge bulk is comparable in area to the Earth. It is a storm that has persisted for at least 300 years. Since the spot's visibility changes dramatically, you cannot be guaranteed a good view every season. In the early 1970s its color began to fade, so that sometimes its northern edge would appear as a colorless shape, and at other times it would seem entirely invisible. The Voyager pictures seemed to show an obvious red spot, but those images were artificially color-enhanced. If you have trouble seeing the spot, a blue (Wratten 80A or 82A) or green (Wratten 58) filter may improve the contrast so that you see it better.

Some notes on the belts and zones:

The *equatorial zone* is the light colored central band of Jupiter. White, orange or yellow, it often appears overrun by festoons and bridges from the nearby belts.

The *north and south equatorial belts* are the dark bands surrounding the equatorial zone. These are the planet's most complex features, and on a night of good seeing the details within them can be most striking. The South Belt is typically the more active, consisting of two sections separated by a thin *south equatorial belt zone*. The area's most interesting activity is the *south equatorial belt disturbance*, a storm that has appeared at irregular intervals.

The storm typically begins with the appearance of a bright spot. A few nights later, a small dark spot forms and eventually spreads into a tortured filament, punctuated sometimes with small knots and small white ovals. As the storm evolves it can spread over much of the south equatorial belt and to the rest of the planet. The storm reaches its greatest intensity after several weeks and then gradually subsides.

Farther from the busy equator, the *temperate belts* are usually not quite as prominent as the features near the equator, but they do offer a rich variety of detail if observed under good conditions with a 20 cm reflector. Sometimes the *north temperate belt* divides horizontally into two parts, often with festoons and a bridge or two crossing over the narrow zone that divides them. The *south temperate belt* can occasionally get even more pronounced than its equatorial neighbors, although normally it is narrower and less well defined.

White oval markings occasionally appear in the *south temperate zone* and *south temperate belt*. A set of three bright oval markings in or near the south temperate belt has endured from 1939 to the present time. It appears that they were not on the planet in earlier years, as observers with the fine refractors did not notice them.

Depending on the seeing, the *north polar region* and *south polar region* may be rather difficult to detect at all, or they could be quite prominent. In most drawings these areas get short-changed somewhat since they are rarely as exciting as the other regions.

## 9.4    Drawing Jupiter

Why draw? There are two reasons: drawing teaches you to see and learn more, and the data you and other observers collect provide planetary scientists with clues to the changing characteristics of Jupiter's atmosphere. The Voyager probes observed Jupiter at close range, but only briefly. Ground-based observations by experienced amateurs, covering decades of time with instruments of approximately even quality, provide a useful picture of another kind.

### 9.4.1    Full disk drawings

A Jupiter drawing is an activity you build up to gradually. First, your telescope is set up, the drive humming, eyepieces selected so that you have a

Figure 9.1. Jupiter, July 29, 1989, at 05:40 EDT. This sketch, made using Harvard College Observatory's 23 cm refractor at 274×, illustrates a fading of the south equatorial belt. Drawing by Stephen James O'Meara.

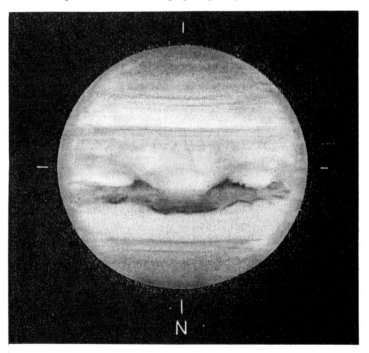

power high enough to see detail but not so high as to magnify the trouble-some effects of the atmosphere of our own planet.

Next, just look. For 10 or 15 minutes, simply direct your gaze at Jupiter. At this stage leave your pencil and paper alone. Just get to know what's there. Gaze across the planet and examine as many details as you can. Try to figure out an approach to your sketch. What features will you want to draw first? What feature is darkest? lightest? How will you represent those various levels of shading in your sketch?

Once your eyes have become full of a cornucopia of Jovian detail you will begin wondering how you will get it all down on paper! The idea now is to record as accurately and as completely as possible. Use a dim, *white* light for Jupiter work, not a red one. Red casts the wrong hue on your paper, and your drawing should reflect the proper relative intensities of the Jovian features. Also, the best drawing pencil is 2B, a soft lead that is popular for sketching. An artist's stub helps to tone the sharp pencil lines into smooth feature reproductions.

Since Jupiter has an oval, flattened shape, thanks to Jupiter's rapid rotation period, use a slightly oval outline. Begin your drawing by sketching the positions of all the major belts, beginning on the west limb, where features will soon be disappearing off the disk. Then, noting the time to the nearest minute, draw the shapes of the details of the belts with the most features, complete with light spots and festoons. Try to keep this phase to 10 minutes, since Jupiter's rapid rotation will interfere. This may seem an insurmountable task at first, but just remember that, as with the lunar drawings, (see Chapter 7) your purpose is not so much artistic excellence as faithful reproduction of what you see on a distant planet.

The fine details within each shape can be added in a more leisurely way (after the 10 minutes) as they will change much more slowly than do their positions. With experience, your answer to the question of what to draw first may become more obvious; usually you first complete the equatorial belts and work into the temperate and finally the polar regions.

Finally, try looking for subtle shadings and details. The more you study the planet, the more you will see, and now that your drawing is almost done, new details that eluded your eye will appear more obvious. 'How could I have missed that?' you might ask about some shading in one of the polar regions.

By this time your drawing should look like what appears on the planet, although some things are wrong. The belts are outlined in heavy pencil, which does not really resemble their delicate borders as they actually appear. Also the pencil lines are too sharp. The solution is to soften the edges of each feature by using the pencil eraser. Then, gently shade the features using your artist's stub, or at least some tissue mounted around the pencil's eraser end. As you proceed with this pleasant task you will see your drawing take on a much more realistic and beautiful look. If you see some very bright spots or oval markings, outline them with dashed or dotted lines.

### 9.4.2    Specific regions

You might want to try drawings of specific regions of the planet, like the details of a very active belt. By waiting a few hours and continuing the same strip you can get a stretch of belt detail on more than half the planet in a single night. Repetitions of this can also record the change of activity on that belt from one week to another, thus helping to acquire important archival records of the planet's changing face.

**Color renderings** Like the details themselves, different colors appear subtly, not dramatically like traffic lights. The larger your telescope, the more distinct color will appear; if a good 7 cm telescope will show the planet as a dull grey or brown, a 15 cm starts to show some reds, and a 20 cm shows other colors more obviously.

When you draw, don't use colored pencils; it is unlikely that you will find a color to match precisely what you see. Simply note the colors with these abbreviations, suggested by the Association of Lunar and Planetary Observers:

> W=white
> Y=yellow
> Y-W=yellowish-white
> Bl=blue
> G=grey
> Bl-G=bluish-grey
> O=orange
> Oc=ochre
> R=red
> t=tan
> Br=brown
> R-Br=reddish-brown

**Intensity estimates** How bright or dark are the various markings on Jupiter? The ALPO intensity scale can simplify your effort to gauge the relative brightnesses of Jovian features:

| | | | |
|---|---|---|---|
| 10 | unusually brilliant | 5 | dull |
| 9 | extremely bright | 4 | dusky |
| 8 | very bright | 3 | dark |
| 7 | bright | 2 | very dark |
| 6 | slightly shaded | 1 | extremely dark |

0    black

Admittedly this scale is somewhat subjective. Very bright zones would normally get 9 or 8, while ordinary zones would range from 8 to 6. Dull or somewhat shaded zones get 6 or 5. The uncertain polar regions are 4, while

3 and 2 generally refer to the darkness of belts. The most intense of dark concentrations deserves a 1, and 0 is assigned only to the dotlike shadow of a satellite.

**Central meridian transits** A planet's central meridian is an imagined line running from the Earth-turned pole through the center. Because the various belts and features do not rotate at the same rates, timing the moment when a Jovian feature crosses the central meridian can be a valuable contribution, because we might understand better the relation of the various belts and zones of Jupiter's atmosphere. A series of accurate timings of features at their local midday would give a continuing record of the rotation periods of the various regions on the planet.

First, you need an accurate clock, set to one of the standard time signal stations like the American WWV or Canada's CHU. A small variety of radio called a 'time cube' is set to receive such signals. You can time the transits of features like the Red Spot as they cross the central meridian, in the same way that nineteenth century astronomer Arthur S. Williams did when he first devised the technique of recording the time of a feature's transit, along with enough description of the feature so that people studying your data will be able to identify what feature you are timing.

Figure 9.2. Jupiter, January 6, 1990. Photographed by Don Parker through a 41 cm f/6 Newtonian reflector, 1.5 s exposure, no filter, on hypered Technical Pan 2415. Rodinal Developer. 'Preceding' means 'west' in our sky.

South

Preceding

The Red Spot would make an ideal central meridian transit timing. However, its oval shape and size can vary from season to season, and thus a single timing is not enough. Make timings as the spot's preceding end, its center, and its following end transit the central meridian.

Once you get proficient at making these transit timings, try doing a few before you begin a drawing. This way your sketch will be much more accurate since you will have paid careful attention to the moments when selected features have crossed the meridian. After your drawing you could make further timings. By checking when the features you drew crossed the meridian, you have tried to verify the accuracy of your drawing.

## 9.5    The Galilean satellites

Although I have seen Ganymede as a perceptible disk through a good 7.5 cm refractor on a night with unusually good seeing, observers normally need at least a 15 cm instrument to detect this. The seeing needs to be very good. Details on Ganymede and Callisto may even appear, although seeing and confirming them is extremely difficult. The best time to look for these details is when the satellites are crossing Jupiter's face and their glare is consequently reduced.

When a satellite crosses Jupiter's face, it could be visible as a sharply defined bright spot. The visibility of a satellite in transit depends on the brightness of both the satellite itself and the Jovian background on which it is projected. Io and Ganymede are bright near the limb, or edge, of Jupiter, but dark near the center. Callisto is dark during almost all of its transits, being often taken for a shadow. Europa is bright near the limb of Jupiter and usually invisible near the central meridian.

When the Earth and Sun are in the orbital planes of the four bright satellites (a situation occurring every six years) mutual events between two satellites take place. These may include the shadow of one satellite crossing the face of another, causing it slowly to disappear (a mutual eclipse), or the passing of one satellite directly in front of another (a mutual occultation).

# 10    Saturn

*Annulo cingitur, tennui, plano, nusquam coherente, ad eclipticam inclinato.*

'It is surrounded by a thin, flat, ring, nowhere touching, inclined to the ecliptic.' One of the most astonishing sentences ever written in science. Christiaan Huygens didn't actually write these words in 1656; instead he wrote an anagram that consisted of a certain number of repetitions of each letter of the alphabet. They deciphered to the Latin words that announced Saturn's ring.

Until that time, Saturn's strange shape had been an enigma. Named after the Roman god of the harvest, its symbol recalls Saturn's scythe. Galileo noticed a pair of 'ears' that seemed to be appended to each side of the planet, but his explanation lacked the self-confidence he had displayed with the moons of Jupiter, the phases of Venus, and the sunspots. The indistinct ears had no obvious explanation. Worse, as a few years went by, they grew smaller and disappeared.

Galileo never did find out about Saturn, never had that first look at Saturn's rings, something that no one should miss. My first experience with a telescope back in 1960 was an awesome surprise. Two bright objects, Jupiter and Saturn, were well placed for this first look through Echo, my 9 cm 'Skyscope'. I remember not being too impressed with the first night's look, seeing only a doughnut shaped light where a planet should be. I learned the most important thing about telescopes that night, that they need to be focused. The next night also was clear, and with the telescope my parents and I were able to adjust the size of the doughnut by pushing the eyepiece in and out. As the doughnut got smaller the image settled into an oval ball, not unlike what Galileo had seen. Then we saw what had always eluded the great seventeenth century observer: the image settled on a small round ball surrounded by an exquisite set of rings.

I was stunned. Having seen the pictures, I knew what Saturn was supposed to look like. We all know what the Pope or a movie star looks like too, but how would you react if they knocked at your door? From the viewpoint of a child with a tiny telescope, that first look at Saturn's rings was unforgettable. In addition to a division between the globe and the rings, there was even a division between two rings! I still wish that everyone could look at those beautiful rings.

## 10.1  Historical perspective

Galileo was not the only one puzzled by Saturn; Gassendi and Riccioli, a Jesuit astronomer, could not understand the changing appearance of this planet either. About half a century after these early looks, Christiaan Huygens solved the riddle, and also discovered a satellite which we now know as Titan. By the 1670s, Cassini had recorded four additional satellites, and his telescope provided such a clear view of the rings that he detected a gap between them. Known as *Cassini's Division*, this gap is easy to see and adds to the beauty of the rings. Early in the nineteenth century Johann Encke discovered a second division, this one in the outermost ring.

William Herschel, the great English observer, advanced Saturn studies by carefully observing the rate of movement of the markings on Saturn's disk. From his observations he deduced a rotation period that differed slightly from belt to belt. He concluded that Saturn had a rather dense atmosphere whose winds carried different belts along at different rates, and that all these rates were fast.

## 10.2 The rings

Saturn's rings are composed of tiny particles of rock or ice. Recent data, mostly from Pioneer and Voyager probes, have rewritten our visual impressions of the three rings referred to as A, B, and C, where A is the outermost ring and C is the thin Crepe ring whose ghostly, almost translucent, appearance reminds us of decorative party paper. Because Saturn's rings are inclined about 27 degrees to its orbital plane, and the planet takes 29 of our years to orbit the Sun, we see the rings at always changing angles. In 1958, the rings' north side was at maximum inclination toward the Sun. In 1966 the rings were edge-on, visible either as a thin line or not at all. Their south side was most highly inclined toward the Sun in 1973, and the rings were edge-on again in 1980. A narrow ring, now known as E, was found in 1966, F was discovered in 1979, and G in 1980.

On July 3, 1989, the north side of the rings was well inclined toward the Sun when Saturn occulted 28 Sagittarii, a sixth magnitude star easily visible through binoculars. For several hours the planet and star put on one of the most wonderful performances ever seen in the sky. As different structures in the rings passed in front, the star faded and brightened, sometimes blinking on and off. For a short time 28 Sgr shone brightly in the Cassini Division between the outermost A and middle B rings. The episode's end was unexpectedly abrupt as the star faded for an instant as the F ring, discovered by Pioneer 11 in 1979, passed in front of it. It was the first time that this elusive ring, located just outside the A ring, could be detected with a back yard telescope!

## 10.3 The globe

In Jupiter's case, the sizes of the features are about the same on both sides of the equator. With Saturn, that is not the case. Let's take a look at these features from south to north, as they might appear.

The southernmost third of the planet is dominated by a dusky band that consists of two parts, the *south polar region* and a wide *south temperate belt.* Unless the seeing is good, all you will see in that region is a combined darkish area. On good nights, however, a narrow, bright, *south temperate zone*, close to the pole, separates the two. The bright *south tropical zone* completes that part of the planet.

Continuing northward, the *equatorial region* begins with a *south equatorial belt,* divided into south and north regions by a very narrow *south equatorial belt zone*. Then there is a wide *south equatorial zone,* a very narrow *equatorial belt,* and a *north equatorial zone*. Depending on the presentation of the rings, one of these zones will likely be obliterated by the rings when you observe. Also, a very dark shadow from the rings is projected on to the ball.

The dark *north equatorial belt* is divided into south and north parts, separated by a *north equatorial belt zone*. Both components of this belt are sometimes narrower than their southern hemisphere counterparts. As we move further north we find that the bright zone areas might be larger and more dominant than in the south, or that the belts are narrower. Here the bright *north tropical zone* and *north temperate zone* are separated by a narrow *north temperate belt*. A brownish *north north temperate belt*, a large whitish *north north temperate zone*, and a *north polar region* that is slightly narrower than the south one complete the major features on Saturn's globe.

This reads like a complicated disk to watch and draw, or an invitation to look through your eyepiece and see some unbelievable detail on the ball of a giant planet. Not true: unless the conditions are very good, you won't see many of these features. Instead you will observe a planet dominated by rings, with a few soft and subtle features on the ball.

In 1964, the Canadian observer Klaus Brasch summarized the details of 55 disk drawings done by observers, in the RASC Montreal Centre newsletter *Skyward*:

> Most prominent during the past few years has been the North Equatorial Belt (NEB) and this year was no exception, as this belt appeared on 51 of the 55 drawings. The dusky North Polar Region was noted on only 18 occasions . . .
>
> Other disk detail drawn periodically was a band corresponding to the North Temperate band, as well as some scattered detail in the hemisphere south of the rings. These features appeared on 5 and 8 occasions respectively, suggesting that they were probably spurious in nature.

Our description of what Saturn should look like mentioned 19 separate features and the ring shadow. The 55 drawings to which Klaus Brasch refers were done by observers whose experience ranged from none (me) to 15 or more seasons, and therefore they tell us much more about what you can expect from Saturn than some theoretical plot of many features.

## 10.4  Drawing Saturn

Because details on Saturn's disk are less distinct, and because of the complicated and changing geometry of the rings, I have always found Saturn more of a challenge to draw than Jupiter. A drawing is not complete with a standard oval disk, as it would be with Jupiter, because you need to include at least an outline of the rings, and their geometry changes with each observing season as Saturn orbits the Sun. The ALPO supplies a series of Saturn forms that keep up with the rings' changing perspective.

### 10.4.1  A cloudy night experiment for clubs

Find a photograph of Saturn (taken by a ground-based telescope, not a spacecraft), in some text book and photograph it so that it can be projected as a slide. Project this slide at one of your meetings, and have each member try to reproduce the slide as a drawing. You may learn some interesting things about how observers tend to overemphasize unusually bright or dark features, like the shadow of a satellite, and how some observers may consistently put the features too far north or south. An advantage of this exercise is that everyone sees the same picture, unaffected by seeing, poor tracking or other temporary telescope problems, and can compare each drawing directly to the projected photograph.

### 10.4.2  Actual drawing

The first thing to look for is details about the rings. If you have at least a 10 cm telescope, you should see the division between the outer A and B rings. If the seeing is good, you may also see the inner Crepe or C ring close to the planet. It is obviously less dense than the others.

With that much in mind, the procedures are essentially the same as for Jupiter. The details are much more difficult to see than those on Jupiter. The colors are more subtle, a fact that was confirmed by the Voyagers, and so there are fewer levels of detail to which you can go in trying a drawing.

### 10.4.3  Estimating conspicuousness

An easy way of determining the relative visibility of a planet's features is to estimate their relative conspicuousness. The first step involves choosing the two features which are most and least obvious to detect, and then rate the other features with respect to these. One of the equatorial belts will probably be the easiest feature, hence deserving of the highest rating, with a dim feature near the pole probably getting the lowest rating.

Figure 10.1. Saturn, sketched by Stephen James O'Meara using a 60 mm refractor, on August 24, 1989.

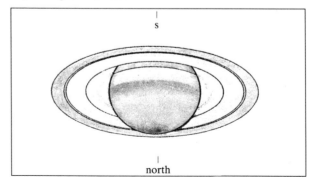

## 10.5  Estimating intensity

Like Jupiter, Saturn has an atmosphere whose belts and zones vary in intensity. In an important sense, estimating intensities is easier with Saturn because we have, with the outer part of ring B, a consistent standard by which we can judge the relative intensities of other Saturnian features. Use an intensity scale of 0 to 10, where 0 represents the rarely seen blackness of a satellite's shadow and 10 is a brilliant white. Ring B's outermost part is usually the brightest and most obviously striking part of Saturn's entire system, and we assign that brightness an arbitrary value of 8.0. At first, try estimating in terms of whole units, so that the dark bands get about a 3, the dusky white zones are 5 or 6, and the equatorial zones might reach 7 or 8. With more experience you may go to half numbers, so that a dark band may get a 2.5. Advanced observers sometimes estimate in tenths, so that different features of a dark band may get a 3.2 or 3.4.

Estimating the intensity of Saturn's features is a way of recording the atmosphere's changing behavior in a quantitative way. These intensity estimates are most easily done in the context of a drawing, although an outline sketch in which features are simply noted by position and then used for intensity estimates is acceptable too. If you do intensity estimates

Figure 10.2. Saturn, April 15, 1985. Photographed by Don Parker through a 32 cm f/6.5 Newtonian reflector, 6 s exposure, no filter, on hypered Technical Pan 2415. Rodinal Developer.

frequently enough, you may find certain features change more frequently or dramatically than others. A record of these changes over a single observing season will document this changing atmosphere. With experience, you will be able to compare features from one season to the next, although the break in between seasons will require a little refamiliarizing when Saturn is observable again.

Like Jupiter, Saturn's global features can change radically, and a whole belt or zone may get more or less intense over a short period of time. Belts have been known to disappear entirely for periods ranging from a few weeks to a season or longer. Rarely, large white spots have appeared, possible storm systems amidst the clouds.

Because Saturn's features are so much less distinct than those of Jupiter, estimating their intensity is both easier and more challenging – easier in the sense that the planet is plainer and its features are easier to distinguish, but harder because they are less distinct.

## 10.6   The moons

A fun trivia question. Try asking someone the names of the nine large moons of Saturn – Mimas, Enceladus, Tethys, Dione, Rhea, Titan, Hyperion, Iapetus, and Phoebe.

Observing Saturn's moons is a different experience from observing those of Jupiter, which are all visible with a small telescope and which are often lined up. With Saturn, the moons are lined up only when Saturn's rings are edge-on. At other times they appear more as a cluster surrounding the planet.

### 10.6.1   Titan

Titan reveals itself through almost any telescope. If you want to view other moons you need at least a 7.5 cm instrument, which will show Rhea and occasionally Iapetus. A slightly larger instrument will let you see three other moons, Enceladus, Tethys, and Dione; 20 cm or larger telescopes are needed for Hyperion and Phoebe, and Mimas, so close to the planet, is harder still. Each year, the *Astronomical Ephemeris*, published jointly in the United States and Great Britain, assists in identifying Saturn's moons by publishing their positions.

Titan is by far the brightest; and with the possible exception of Triton of Neptune, it is the largest satellite in the entire solar system, much larger than Pluto and larger than Mercury. It also has an atmosphere that consists of nitrogen, methane and other hydrocarbons, and traces of argon and hydrogen cyanide – an unappetizing soup that Voyager I tells us is not quite twice as thick as ours is at sea level.

The complex atmosphere of this world reacts, with sunlight as a catalyst, to produce something that is a bit like smog. Our Earth had a similar atmosphere before it grew rich with oxygen. On Titan, the smog is so thick that it precipitates to the surface, where it stays indefinitely. A close look at this material would give us the clearest view yet of what the Earth was like in its early history. Moreover, there may be rivers and oceans of the hydro-carbon called ethane. When the Cassini spacecraft arrives at Titan on some future date, it is to release Huygens, a probe designed to float on this strange sea!

Of course, we don't get to observe any of this with a small telescope here on Earth. With a 30 cm telescope, possibly we can discern a disk. All we can do, then, is to look and wonder at the mystery of this distant moon.

### 10.6.2 Iapetus

Iapetus is special in that its magnitude varies by a large amount – 1.5 magnitude – from 9.5 to 11.0. An advanced project is to try estimating the changing brightness of Iapetus, using Titan as a comparison. (Titan is not entirely safe as a comparison standard; although it is fairly constant in magnitude, its reddish hue makes it hard to compare with less colorful Iapetus. Also, Titan may appear to fade as it closes in on Saturn's globe and gets swamped by its light.)

Because Iapetus is one of Saturn's farthest satellites (eighth in order of distance) it can be far enough from the planet to allow good estimates. These large brightness changes are supposed to be independent of Iapetus's apparent distance from the planet; obviously, estimates made at varying distances from Saturn must be adjusted, since, like Titan, the closer Iapetus gets to Saturn the fainter it will appear because of Saturn's glare.

### 10.6.3 Phoebe

In the days before spacecraft discovered many more moons, Phoebe was a rarity in the solar system; it revolved around its parent planet in a direction opposite to that of the others, or in retrograde motion. Now we know of several other satellites that orbit their parent planets in this way. Astronomers at the time treated it with special, though grudging, respect. Solar system bodies were to orbit in a manner set when the primordial solar system cloud began to rotate before even the Sun was formed. Why should a moon orbit the wrong way? Actually, retrograde orbits are a sign that these moons did not join the Saturn family at its formation but arrived later; they are probably asteroids whose orbits around the Sun coincided with that of Saturn, leading to their capture.

In a whimsical mood, some astronomers posed the question in poetic terms; this poem was published in *The New York Times* of January 22, 1967:

> Phoebe, Phoebe, whirling high
>   In our neatly plotted sky,
> Phoebe listen to my lay,
>   Won't you swirl the other way?

Never mind what God has said,
 We have made a law instead.
Have you never heard of this
 Nebular hypothesis?
It prescribes, in terms exact,
 Just how every star should act.
Tells each little satellite
 Where to go and whirl at night.
And so, my dear, you'd better change;
 Really we can't rearrange
All our charts from Mars to Hebe
 Just to fit a chit like Phoebe.

# 11  Mars

A mind of no mean order would seem to have presided over the system we see, – a mind certainly of considerably more comprehensiveness than that which presides over various departments of our own public works. Party politics, at all events, have had no part in them; for the system is planet wide. Certainly what we see hints at the existence of beings who are in advance of, not behind us, in the journey of life.[5]

Bright points in the sky or a blow on the head will equally cause one to see stars.[6]

Percival Lowell, 1895

What is it about Mars that so captivates us? Its angry red color possibly associated it with 'manhood', hence its symbol. For much of the latter part of the nineteenth century, talk of canals on Mars fueled our conception of what we wanted our neighbor planet to be, that Mars was inhabited by intelligent beings.

Late in the nineteenth century, Schiaparelli was fortunate enough to see Mars through optics good enough that some fine detail was visible. Although he was not the first to notice possible long, straight lines on the surface of Mars, he did publicize the 'Canali' – although a strict translation of this term would mean 'channels', the looser and more exciting 'canals' quickly made it into the lexicon, and 'life' had taken on a meaning of its own on Mars.

If the canals were born through Schiaparelli, they were nurtured to maturity by a very wealthy, eccentric, and first class amateur observer from Massachusetts named Percival Lowell. For the favorable Martian apparition of 1894 he constructed an observatory in Flagstaff, Arizona, for the purpose of studying Mars. The 24 inch (61 cm) refractor was one of the finest in the world at the time, and with it Lowell drew stunning drawings of Mars that

5  P. Lowell, *Mars*, p. 209. London: Longman, Green and Co., 1895.
6  Constance Lubbock (ed.) *The Herschel Chronicle: The life-story of William Herschel and his sister Caroline Herschel*, p. 86. Cambridge University Press, 1933.

showed long canals crisscrossing the planet, forming a huge network that included dark oases where the canals intersected. Some canals were even double, tracing parallel paths across large areas.

Lowell's observations were ridiculed during his life, and it was partly to recover from his disappointment that his work was not accepted that he began a search for 'Planet X' that eventually led to the discovery of Pluto by Clyde Tombaugh in 1930. The two stories of Mars and Pluto would merge unexpectedly in Texas a quarter of a century later.

Through this century, events conspired to keep Martians in the news often enough. In 1938, Orson Welles and the Mercury Theatre presented a realistic and scary radio show called *War of the Worlds* in which Martians were landing in New Jersey and invading the Earth. Even though the program was labeled as fictitious, many listeners, tuning in late, believed that Earth was indeed being invaded. Panic was widespread, even though the astronomers at the 'Mount Jennings Observatory' could not possibly have observed the launch from Mars two days earlier as the story claimed. As Clyde Tombaugh has pointed out, Mars was in conjunction with the Sun at the time and not visible from Earth. (Incidentally, a rebroadcast of the program 50 years later caused another panic among listeners in Portugal.)

The 'flying saucer fifties' also kept talk of Martians alive. Even though most scientists did not take Lowell's observations of Mars too seriously, ground-based observations were not good enough to answer the question of Martian life definitively.

Although most people place the flight of Mariner IV as the first proof against canals, a windy night in 1950 provided evidence much earlier. As Gerard Kuiper and Clyde Tombaugh started to observe Mars with the 82 inch (208 cm) reflector at McDonald Observatory in Texas, a biting and violent wind sprang up from the north. Fortunately Mars was in the southern sky and the dome helped to block out the wind. The seeing was excellent on that night, enabling the observers to see Mars in a way Lowell would have envied. The 'canals' were there, looking like canals because irradiation from bright areas made them look that way. By stopping down the aperture so that the bright areas did not irradiate into the faint darker markings, the 'canals' resolved into separate, small areas that clearly were not canals.

Tombaugh understood that the mistake Lowell had made was simply not using enough focal length in his telescope; had the refractor been just a bit longer, the canals would have resolved into craters and other features, and much of the Martian controversy might have been avoided. Lowell was a good observer, reporting accurately what he saw. However, he did want the canals very badly. A pacifist in his politics, Lowell wanted to point out the advantages of a world united by technology and a will to progress together. The canals and oases he thought he saw on the face of Mars offered evidence of such an advanced race working together to overcome a water-starved environment.

But Martian talk would continue along other lines of 'evidence'. As

Figure 11.1. This 1971 map of Mars is based on visual observations through a 30 cm reflector stopped down to 10 cm, and through a 15 cm refractor, by Michael Mattei. Printed by permission of M. Mattei. Numbered features include: (1) Elysium, (2) Syrtis Major, (3) Hellas, (4) Sabaeus Sinus, (5) Meridiani Sinus, (6) Margaritifer Sinus, (7) Mare Acidalium, (8) Solis Lacus, and (9) Tharsis.

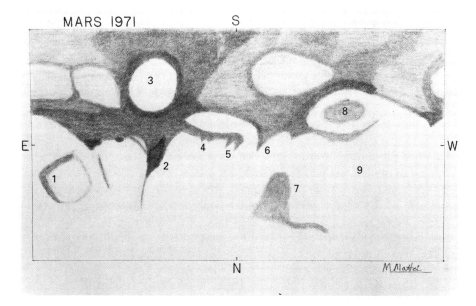

Figure 11.2. This 1973 map of Mars is based on visual observations through a 15 cm refractor by Michael Mattei. Printed by permission of M. Mattei.

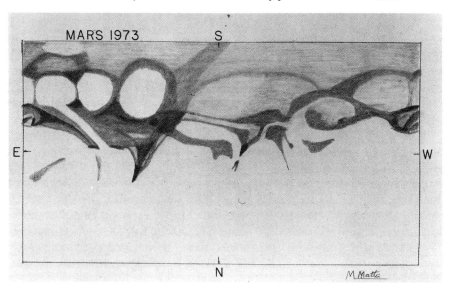

recently as 1959, when Iosef Shklovskii postulated that Phobos was hollow, theories about intelligence on Mars were being developed. Coming just three years after the best apparition in years, this concept of a huge, hollow, orbiting platform was attractive to those who still held out some hope for a civilization there.

## 11.1  Observing Mars

The familiar platitude that Mars is a 'planet of mystery' ignores the real challenge of this enticing planet. This is the planet most like Earth, the world to which we still look longingly for signs of life. It is well seen for only a few months every 26 months, a bit less than half the time for visibility of the other planets. These periods of visibility are known as 'apparitions', and the point at each apparition when Mars is opposite the Sun and in the sky most of the night is called the 'opposition'. These oppositions occur every 26 months, a chance to observe a planet whose red color is so pronounced that it really stands out in a dark sky.

Unless Mars is near opposition, with a disk at least 5 seconds of arc in diameter (and more if you are just beginning), there is not much point in attempting a drawing, for the red planet's features are so delicate that you need all the advantages of closeness to see them clearly. Depending on where Earth and Mars are in their orbits, the apparitions can be favorable or unfavorable. The orbit of Mars is not as circular as that of Earth, and therefore there is a big difference between the apparent size of Mars in oppositions that occur when it is near aphelion (farthest point from the Sun) and perihelion. At perihelic oppositions, also known as favorable ones, Mars can be as close as 56 million kilometers from us, but at aphelic oppositions the distance can exceed 100 million kilometers. Perhaps that is one reason why so many planetary observers 'live' for those rare opportunities to see this planet from a privileged seat.

When you see Mars clearly the detail can be stunning. In 1971 I made drawing after drawing of a large disk whose 24-hour rotation gave a continuous display of familiar features. Unlike those of Jupiter, these features are mostly surficial and not atmospheric, so details should appear to change only slightly with time.

The range of detail that you may see is very large. At any opposition, even with a 5 cm telescope, you should be able to see one of the polar caps. But at least a 15 cm telescope is needed for reliable observation of other features. Mars's features have a huge range of difficulty in observing, from the bright polar caps to the dark markings of Syrtis Major and Solis Lacus, to areas like Mare Acidalium, prominent in some apparitions and not in others, to the most difficult objects like the craters.

The face you see from night to night may not appear to vary by much, since Mars's rotation period is only 40 minutes longer than ours. You do get a significantly different view after about a week, by which time Mars's features have moved eastward by almost five hours.

Whether you can see a feature depends on the size and quality of your telescope, your own experience, the effects of weather and seeing on Earth, and the effects of the weather on Mars. If you observe Mars when it is almost at opposition, it will present a fully illuminated disk. In the months before opposition, Mars shows a gibbous phase, and you see the evening terminator on the east side of the planet. After opposition, the sunrise or morning limb appears on the west side.

## 11.2  Drawing Mars

The descriptions that follow suggest that Mars will explode with detail the minute you look at it. Actually, you may see absolutely nothing on the disk if the seeing is poor, or even if the seeing is good and Mars is undergoing one of its global dust storms. If you are not using high enough magnification, the planet's details might be washed out. You may need to use one or more of the following methods to increase the detail of what you see:

1.  Use the highest power you can get away with; that is, the highest power you can use before the seeing causes the image to deteriorate. Sometimes you will need up to 400 power to avoid the irradiation effect that comes from too much light occupying too small a space. Sometimes the seeing may be so poor that even though the night is sparkling clear, there is no sense in attempting a drawing.

2.  What you see on Mars depends on whether you are looking at surface features or those in the atmosphere. An orange filter like Kodak's Wratten Series no. 21, a light red filter like Wratten 23, and a red filter like Wratten 25 will increase the contrast between the bright desert features and the darker maria areas. These filters will act to reduce the total light from Mars that reaches you, and at the same time most of the light it does accept will enhance the surface features, so that you will identify them more easily. Also, a neutral-density filter might reduce the contrast against a dark sky.

3.  With a telescope greater than 30 cm aperture, an aperture mask made of cardboard can go in front of the tube. If you use, for example, a 40 cm f/5 reflector, you may want to cut a hole in the cardboard that renders your telescope a 20 cm. The focal length will be the same, but the f/ratio will be 10, considerably longer. On nights of marginal seeing, this procedure often improves the image quality. Also, offset the 13 cm hole so that not much of the secondary mirror appears. The resulting light path will be clearer, permitting more contrast. For telescopes smaller than about 30 cm, aperture stops include so much of the secondary mirror that the image quality deteriorates.

Before you begin your drawing, just look. Get to know the planet, where the polar cap is, on what side the terminator is, and where the major features are. Compare what you see to what is on the map.

And now, the drawing: The Association of Lunar and Planetary Observers recommends that you use a drawing outline disk at least 42 mm in diameter; otherwise there will not be sufficient room to record detail accurately. Start by drawing the terminator, the position of the polar cap, and then outline the major surface features. Outline these features lightly at first; you can shade them later. Do not spend more than five or ten minutes on this step. As you continue to see more detail and get a clearer impression of where everything is, you will probably erase some of your earlier marks. Write down the time of the midpoint of this part of your drawing.

Darken the appropriate features, using an artist's stub to blur your pencil lines. As you have looked at Mars you have gradually become aware of other features. Now is the time to sketch in these finer details.

Finally, compare your drawing to a last look at the planet. Now relax; you have finished!

## 11.3  Kinds of changes to expect

The features you see can undergo three types of changes: secular, seasonal, and atmospheric.

*Secular changes* refer to long-term changes in the appearance of certain features. From apparition to apparition, features as important as Syrtis Major and Solis Lacus have shown significant changes.

*Seasonal changes* involve what you would expect from a planet that in some ways is very much like the Earth; the increase or decrease in size of the polar caps and the changes in contrast and colors of certain topographic features.

We could follow the Martian seasons better if we had a calendar. Its year of 687 days, eccentric orbit, and axial tilt of about 25 degrees, give Mars seasons that differ from ours in significant ways. They are longer, and northern hemisphere spring and summer are longer than autumn and winter.

With this information observers have found it convenient to use a Martian date calendar that has 12 Martian months with 365 Martian days. In a way this has become imaginary since Mars has almost twice that number of days in its year, but setting up a calendar like this enables us to understand seasonal changes on Mars in Earthly terms. We determine the Martian date with the aid of a value called $L_S$ or longitude of the Sun along the ecliptic as viewed by an observer on Mars. The Martian vernal equinox is where $L_S$ is defined as 0 degrees. $L_S$ increases through 360 degrees, with 90, 180, and 270 degrees corresponding to a summer solstice, autumnal equinox, and winter solstice. From this value, the Martian date has been calculated and is published each year in the *ALPO Solar System Ephemeris*.

*Atmospheric changes* involve yellow clouds that are dust storms, white and blue clouds, and morning frost or fog. The most dramatic dust storms tend to occur just after Mars reaches perihelion, which means that often during the best perihelic apparitions we see their effects. These storms can cover the entire planet, making its surface details invisible.

One night in the late summer of 1971 I noticed that part of a familiar feature was simply missing; an amorphous yellowish spot had covered it! Later that year I could see no detail at all. This was the 1971 Mars dust storm, a planet-wide event that interfered with the early work of the Mariner 9 spacecraft.

## 11.4   Surface features

The most obvious features on Mars are the polar caps, the dark 'maria', or the brighter desert features; all also known as 'albedo features'. These features darken in a progressive pattern that stretches from pole to equator as spring evolves into summer on a particular hemisphere.

*Solis Lacus* is informally known as the 'Eye of Mars'. When this feature is near the central meridian it is fairly obvious. Its normal appearance has a similarity to the pupil of an eye, although sometimes the effect is not that definite. Just north of Solis Lacus is a feature called *Tithonius Lacus*, and 'flowing' eastward from this 'lake' is a very subtle channel called *Coporates*. These two features were identified in spacecraft images as a huge canyon thousands of kilometers long. It has been given the name *Valis Marineris* – an ultra-grand canyon about the length of the continental United States, but still not identifiable as a canyon through Earth-based telescopes.

*Syrtis Major* is another highly prominent feature, a plain whose eastern edge changes significantly with the seasons. Sometimes this edge is sharp

Figure 11.3. Mars, sketched on February 11, 1980, using Harvard College Observatory's 23 cm refractor, by Stephen James O'Meara. South up.

and straight, and at other times it is fuzzy and more curved. Meanwhile, smaller features nearby undergo longer term secular strengthenings and near disappearances.

*Elysium* is a bright orange region that spacecraft have identified as heavily volcanic. Not far from Elysium, a dark streak called the Aetheria Darkening, also known as the Hyblaeus Extension, was reobserved in 1978 after a long absence. At certain times of the Martian year look for bright temporary features that appear in Elysium. Also, on occasion a small section called Albor sports a whitish cloud.

*Tharsis* is a volcanic area that includes the gigantic spacecraft-observed Olympus Mons, the largest volcano yet observed in the solar system. Like the Hawaiian volcanoes on Earth, Olympus Mons has grown over a huge span of time. The Tharsis volcanoes are an order of magnitude larger than anything on Earth. On Earth continental plates shift about, thereby allowing active volcanic areas to appear to move from place to place. The Martian crust is apparently much more stable for much longer periods of time, allowing volcanoes to grow to monstrous sizes.

There are really two sets of names for many Martian features; for example the classical name for much of the Olympus Mons mountain is Nix Olympica. Spacecraft images so definitely identified the nature of so many features that renaming them seemed appropriate. However, since many of the renamed features are too difficult for visual observation, visual observers have tended to stay with the old, classical names. We observe Nix Olympica from the ground, but were we to visit Mars we would climb Olympus Mons. In any event, an attempt to climb this mountain would not be terribly stressful, since the grade is very gentle. However, since the mountain occupies an area about the size of Arizona, it would be a long walk!

*Hellas* is a southern hemisphere feature that may be obscured by haze in late autumn and winter, clearing again in spring. In summer the feature darkens to a dark yellow.

Features formerly thought to be canals or channels are detectable by experienced observers. The Ganges 'Canal' was not difficult to see during the 1986 apparition, although at other apparitions it is invisible. Occasionally it shows as a fine line, and then you may even detect is as double, just as Lowell did. At other times it is too broad to be considered canal-like. This change of 'canal' visibility is typical, being prominent in one apparition and absent in another.

*Craters* are extremely difficult objects to detect, even by experienced observers. Two of the largest are Huygens and Schiaparelli. Almost 500 kilometers each in diameter, these are very large craters and should be visible if the seeing permits it. Huygens is just south of the Syrtis Major, and appears as a tiny circular feature that is dark unless it is frost covered, in which event it is light colored. Crater Schiaparelli is different; during the apparition of 1986 Arizona amateur Gary Rosenbaum, among others, observed it as white with frost.

*Polar regions* are complex, with seasonal white caps that are among the easiest Martian features to see. There are also atmospheric polar 'hoods' that consist of crystals of frozen water and carbon dioxide and appear as greyish patches.

During the darkest part of winter in the northern hemisphere, the polar region is dominated by the large polar hood or haze. With the advent of spring, the hood retreats until it is gone, exposing the brighter polar cap. Then in late spring the hood reappears very suddenly in an event known as the 'aphelic chill' as water vapor condenses.

Sometimes the hood is so bright that it cannot be distinguished from a cap; you might be able to tell the difference by using a red filter which might make the hood apparently vanish, showing a smaller, sharp-edged cap underneath the hood. A somewhat rare feature known as the *Rima Tenuis* cuts through the cap as a small rift. In 1984 it reappeared after a very long absence since 1918.

The south polar region behaves differently. There is certainly no effect like the aphelic chill. In early southern hemisphere summer the haze dissipates, leaving a bright south polar cap which then breaks up as it melts; sometimes you can see bright detached sections. A smaller feature known as the 'Mountains of Mitchell' (actually just a plain) juxtaposes with the south polar cap around midsummer.

### 11.4.1  The atmosphere

Mars has a dynamic atmosphere whose effects are exciting to observe from Earth. We can explore these through a discussion of the types of clouds that can be seen:

*White orographic clouds* consist of water vapor and appear over the great volcanic regions like Elysium and Tharsis. The clouds appear during the spring and summer in these areas, beginning around local noon. The Elysium cloud can be as bright as the nearby north polar cap. If you see these clouds just as they are forming, watch them closely as they may develop quickly over the course of an hour or two. In the Tharsis region, 'W clouds' can form and are best seen with the aid of blue or violet filters (Wratten 38 with telescopes under 25 cm, Wratten 47 with larger telescopes). Like the orographic clouds that develop around mountains here on Earth, these clouds probably form as air flows up and across these mountains.

*Yellow clouds* can form and disappear over specific parts of the planet. These are the beginnings of dust storms which eventually can obscure major features or – as in 1971 – the entire planet. Because these clouds are the result of dust particles carried high into the atmosphere by rapidly increasing winds, they tend to form very quickly and take their time dissipating. Since these are yellow clouds, use a Wratten 8, 12 or 15 yellow filter to emphasize them. Since the Solis Lacus is one of the areas that is prone to these clouds, it deserves careful watch. These clouds also appear less frequently northwest of the Hellas and Chryse basins.

*Limb haze* brightens the sunrise or sunset limb as a mist best seen with a Wratten 38 violet filter on a 25 cm or smaller telescope, or Wratten 47 with larger telescopes. They could contain different types of atmospheric particles, including $CO_2$ and water vapor.

Also, a somewhat rare event is the 'Syrtis blue cloud', a bluish haze of varying intensity that can overlay Syrtis Major. When this delicate cloud is present, Syrtis may be dimmed or even obliterated. The cloud is best seen near sunset on Syrtis.

Fogs and frosts occur frequently as bright patches and are difficult to separate from white clouds, since both effects appear much the same. If the feature is brighter using a blue filter than a green one, then it is a cloud. Otherwise, it is fog or frost on the surface.

## 11.5  Phobos and Deimos

Although few people have observed these moons, they are not really that faint. Phobos is about 12th magnitude, and Deimos is about a magnitude fainter, but somewhat easier to see since it is usually farther from the planet. The reason both are so difficult is that they are faint relative to the great brightness of Mars.

Their story begins in fiction, with the writings of the seventeenth century English writer Jonathan Swift. In the third book of his classic *Gulliver's Travels*, Swift wrote about the Laputians, unusual people who live in an 'island in the air' that flies at no more than a few kilometers above the ground. The piece is intended as a satire against the Royal Academy of Great Britain, whose scientists Swift accused of ignoring the concerns of the people – an early example of the tradition of anti-science literature that unfortunately continues to be popular today. Since *Gulliver's Travels* was published in 1726, 151 years before the moons' actual discovery, the following announcement of the existence of two Martian moons is remarkable:

> They have likewise discovered two lesser Stars, or *Satellites*, which revolve about *Mars*; whereof the innermost is distant from the Center of the primary Planet exactly three of his Diameters, and the outermost five; the former revolves in the Space of ten Hours, and the latter in Twenty-one and an Half; so that the Squares of their periodical Times, are very near in the same Proportion with the Cubes of their distance from the Center of *Mars*; which evidently shews them to be governed by the same Law of Gravitation, that influences the other heavenly Bodies.[7]

During the favorable opposition of 1877, Asaph Hall, using the fine Clark refractor of the US Naval Observatory, began a search for a Martian satellite. His discoveries of two moons came as a surprise, since searches at an earlier favorable opposition, 1862, had also been made with high quality telescopes and nothing had been found.

---

7 J. Swift, *A Selection of His Works*, ed. P. Pinkus, 1726, p. 162. Toronto: Macmillan, 1965.

It is a compliment to Hall's persistence that his 1877 search was doubly successful. In fact, years later Russian astronomer Felix Zigel tried to explain the failure to find the satellites earlier as evidence that they did not at the time exist, and that by 1877 a Martian civilization had built them as spacecraft!

Figure 11.4. Mars, June 24, 1986. Photographed by Don Parker through a 32 cm f/6.5 Newtonian reflector, 5 s exposure, no filter, on hypered Technical Pan 2415. Rodinal Developer.

South

Preceding

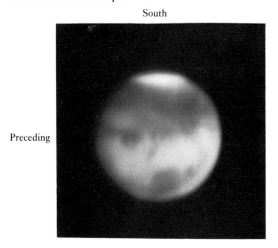

Figure 11.5. Mars, September 18, 1988. Photographed by Don Parker through a 41 cm f/6 Newtonian reflector, 3 s exposure, no filter, on hypered Technical Pan 2415. Rodinal Developer.

South

Preceding

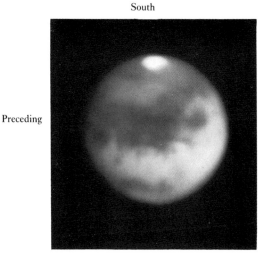

The best way to observe the moons is to obtain an ephemeris that shows their most favorable elongations from the planet; i.e. the times when they are farthest from the planet and most easily seen. (The *Astronomical Ephemeris*, published annually, offers such a schedule.) Your times need to be accurate as the moons will not be favorably placed for more than an hour or two at a time, so fast are their periods of revolution around Mars. Then, with Mars out of the field of view at high power, look carefully for the faint satellite. An occulting bar or spot mounted at the telescope's focal point (much the same way eyepiece crosshairs are mounted) could also block the light of Mars so that the moons could be more easily seen.

## 11.6  Mars thought

The legendary dedication of Mars observers comes partly because favorable oppositions are so rare. In order to get a better view, Percival Lowell even had his entire telescope, mounting, and observatory moved to Mexico for part of the 1894 opposition.

The work of serious Mars observers contrasts with someone who once submitted a Mars 'drawing' that consisted of a roughly hewn circle with a dark splotch in the middle. 'Is that the Solis Lacus?' the observer was asked. 'Oh no,' came the reply, 'that's the mark left by the compass that drew the circle!' Mars offers more than that. Its many changes make it not a light in the sky but a real place whose maximum temperature of 20°C is comfortably warm, although the atmosphere is so thin that this warmth exists only to a few inches from the surface. It is a place with violent storms and bitterly cold nights. Still, of all the objects we see in the sky at night, Mars is the most like home.

# 12   Five planets worth watching

At most astronomy club meetings you will learn much more about Jupiter, Saturn, and Mars, than about any of the other planets. These 'big three' offer so much to a visual observer that they do indeed monopolize most of the planetary observing time when they are in the sky.

This chapter considers the five other planets, in order of ease of observing: Venus, Mercury, Uranus, Neptune, and Pluto. They lack the details, either surface or atmospheric, that make the other three so complex, but all five are worth watching.

## 12.1  Venus

By far the brightest of all the planets, Venus usually dominates the early evening or morning sky. Although the Roman goddess Venus was hermaphroditic, the planetary symbol is an ancient one representing female fertility. Venus is so prominent when not in conjunction with the Sun that one would think that telescopic examination would reveal an enormous amount of detail. Not so: because the planet's surface is completely obscured by a thick cloud, ordinary telescopes are next to useless in revealing anything about it. Aside from occasional darkenings in her clouds, Venus does not usually reward an attempt to sketch. The change in phase is easy to see, however. Like the Moon, Venus undergoes changes from crescent to full phase, but, unlike the Moon, it is brightest when in crescent phase and faintest when full. It is easy to understand this. When Venus's phase is full, we see all of its sunlit face, which means that it must be on the opposite side of the Sun from us and at its most distant. As it approaches our side of the Sun, its phase shrinks toward a crescent and it appears larger.

Because it is a clear example of a 'greenhouse effect' out of control and dominating the planet, it is vital that we learn as much as we can about Venus. However, the omnipresent shroud really relegates the planet to study by spacecraft. It is only slightly smaller than Earth, and it averages about two thirds Earth's distance from the Sun. But since Mariner II arrived there on December 14, 1962, we have recognized that its surface temperature runs at almost 450°C, almost as hot as Mercury which is half Venus's solar distance! It is obvious that the thick cloud is responsible for keeping sunlight from escaping the surface, thus causing a most severe 'greenhouse effect' in which the planetary temperature simply rises and rises. With the same effects being feared for Earth as a result of increasing carbon dioxide in our atmosphere, it makes sense to study Venus as intensively as we can.

I remember sitting in a high school classroom that day in December when Mariner's data streamed back at us as a Christmas present. In my naive way, I thought that the systems might not switch on after their long trip from Earth, and in fact there was a minor problem to get the craft to transmit its data. But as announcements described the data that were streaming in, a friend of mine wondered how nice it would be 'if we could only get Mariner back to Earth and put it into a museum, the first spacecraft that really went to another planet!'

### 12.1.1  Observing Venus

The phases of Venus are so obvious, especially when less than half, that you can see them even with a pair of powerful binoculars. Some sharp-eyed watchers have even reported a crescent phase with naked eye.

Venus is so bright that it is observable in daylight for much of the year. It is very difficult to find, but once you find it you will be surprised at how bright it is. The best way to see Venus in daylight is first to see it in the evening or morning sky and estimate its angular distance east or west from the Sun. If Venus is in the morning sky, compare its angular distance from the Sun. Later in the day you can try to find it again, keeping it in sight later using its position relative to terrestrial objects.

In daylight it is easier to find the planet with a pair of binoculars which you can move much more quickly than the finder of a telescope. Once you have seen Venus, then find it in the telescope's finder, and then you are ready for a telescopic look. Another way is simply to set your telescope's equatorial mount at Venus's declination and sweep a bit with the telescope.

## 12.1.2 Advanced observing

It is almost impossible to see any details on Venus, including its cloud structure. Some observers, particularly Lowell, have seen a system of planet-wide 'spokes' – now accepted as an optical illusion that is not real.

Figure 12.1. (a) Venus, drawn on July 1, 1982, at 11:00 EDT using Harvard College Observatory's 23 cm f/12 refractor at 360×. Seeing 9. Drawing by Stephen James O'Meara. (b), (c) Venus, June 6 and 7, 1983.

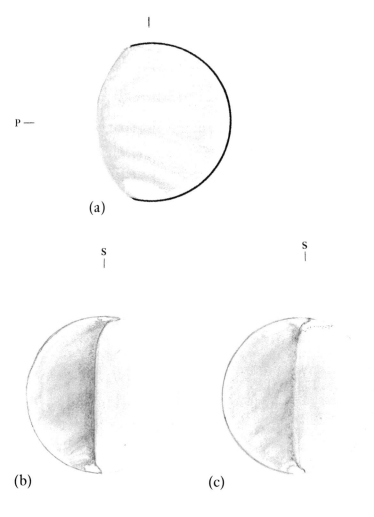

*Terminator shadows* are a relatively common feature best seen when the planet is at half phase. On the terminator there appears a dark shadow that gradually gets brighter toward the planet's edge or limb. Even through a 6 cm refractor, you might see various shades of grey. This feature is most prominent when Venus is in a dark sky.

Occasionally *dark clouds* angle out like tiny fingers from the terminator. They appear to be almost projected from the terminator, and parallel to each other.

Other faint and subtle cloud markings can occasionally be observed in daylight. These features are very difficult to detect, however.

A brightening of the ends of the horns or cusps are most prominent when the planet is in crescent phase. The effect is known as *cusp caps*. Although this may be a real effect caused by foreshortened cloud bands, it is just as likely an optically induced effect involving contrast between the cusps and the surrounding dark sky.

As the planet approaches inferior conjunction, when it is closest in angular distance from the Sun, the cusp extensions get longer and longer until, near conjunction, a *ring* of light surrounds the otherwise invisible planet. This is a Venusian atmospheric effect caused by sunlight scattered through clouds along our line of vision. It is more common to see an almost complete ring of 320 or 340 degrees, than a full 360 degree ring.

**Warning: Be careful of the Sun which is nearby; don't let its light accidentally fall down the mirror. Blocking the Sun by having someone stand near the front of the tube in the Sun's direction, perhaps also holding a piece of cardboard in front of the telescope just to cast a shadow, will increase your chances of a successful observation.**

### 12.1.3  Ashen light

When Venus is at half phase or less, you should try observing the Ashen light, a faint glow that makes Venus's unilluminated face visible. First detected as early as 1643 by Riccioli, a Jesuit astronomer, the Ashen light is unlike the earthshine on the Moon, which is clearly caused by sunlight reflected off Earth. Venus has no large satellite to cause this effect, and the cause of the Ashen light is unknown. Like all the other Venusian details, this effect is usually difficult to see, although occasionally it is as bright as the earthshine on a crescent Moon. Episodes of bright Ashen light can last up to several days.

### 12.1.4  Transits

Transits of Venus occur when that planet passes between Earth and the Sun. These events occur in pairs and are extremely rare, the last two pairs having occurred on June 6, 1761, and June 3–4, 1769; December 9,

1874 and December 6, 1882. The next transit season is approaching; its two events will occur on June 8, 2004, and June 6, 2012, the June 8 event being partially visible from eastern North America, and fully from Europe. These transits occur in very narrow intervals of our calendar, in four day windows centered on June 7 or December 9. They occur so infrequently because Venus's orbit is inclined at about 3.5 degrees to the ecliptic; at a greater distance from the Sun than Mercury, the alignments do not occur nearly as often.

Some observers have reported seeing a 'black drop' effect. In 1882, this optical illusion appeared as the limb of Venus stretched towards that of the Sun, making the planet's shadow look like a black teardrop for a few seconds. Also, a bright circle of light around Venus has been reported when Venus is in partial transit.

## 12.2 Mercury

The name Mercury recalls its fast motion, hurtling through the sky with its mythical staff caduceus.

The trick to Mercury is to see it! The nearest planet to the Sun, Mercury's motions are rapid, its position in the sky changing from night to night. The problem has nothing to do with Mercury's magnitude; in fact when it is visible it is one of the sky's brightest objects. It is its always-small elongation (angular distance) from the Sun that makes it tricky to spot. Western (morning) and eastern (evening) elongations come and go so quickly that you really have to plan an observing program to catch it. Many professional astronomers have never seen Mercury. Can you?

Figure 12.2. Mercury, drawn on July 1, 1982 at 11:40 EDT using Harvard College Observatory's 23 cm refractor. Drawing by Stephen James O'Meara.

| South

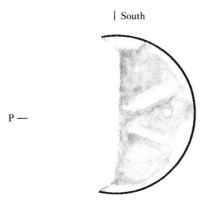

### 12.2.1 Observing Mercury

Like Venus, Mercury has phases, and also like Venus, this planet is easiest to see when it appears as a slender crescent. Unfortunately it is never high in the sky at night, so the seeing in our own atmosphere rarely permits a steady view. Mercury's color appears to vary from observer to observer. Some report a yellowish tint, others report a rose color.

One thing you can look for is a much 'softer' southern horn than a northern one, an effect observed by many observers for years. When the seeing allows it, you may notice faint dark markings and sometimes bright streaky effects also. Although some markings can be seen with apertures as small as 10 cm, a 20 cm or larger telescope is preferred to see any detail on this difficult planet.

Like Venus, Mercury does transit the Sun, but these events are not nearly as rare, some 13 transits taking place per century. A $5\frac{1}{2}$ hour event occurred on November 10, 1973, and on November 13, 1986. Another will happen on November 6, 1993.

## 12.3  How the outer planets were discovered

Our discussion now turns to a new type of planet, not from physical structure or orbit, but from our own historical perspective. The three outer-most planets are interesting to see, if you can find them, but their stories are something special. I remember sitting at dinner one night, listening spell-bound as my father recounted the story of how these planets were found.

### 12.3.1  Discovery I: Uranus

Until 1781, the five major planets were observable, obvious, and complete, and aside from the 'Bode's law' inconsistency that seemed to suggest something between Mars and Jupiter, the solar system's structure was apparently well understood.

Late in the evening of Thursday, March 13, 1781, only eight years after he had begun his work with the stars, William Herschel was observing the region of H Geminorum as part of a program to observe and record as many objects as possible:

> I perceived one that appeared visibly larger than the rest; being struck with its uncommon magnitude, I compared it to H Geminorum and the small star in the quartile between Auriga and Gemini, and finding it so much larger than either of them, suspected it to be a comet.[8]

---

8  W. Herschel, 'Account of a comet', in *The Scientific Papers of Sir William Herschel*, ed. J. L. E. Dreyer. London: The Royal Society and the Royal Astronomical Society, 1912.

News of this 'comet' spread rapidly. Charles Messier wrote of his incredulity:

> I am constantly astonished at this comet, which has none of the distinctive characters of comets, as it does not resemble any one of those I have observed, whose number is eighteen.
>
> I have since learnt by a letter from London that it is to you, Sir, that we owe this discovery. It does you the more honour, as nothing could be more difficult than to recognize it, and I cannot conceive how you were able to return several times to this star – or comet – as it was absolutely necessary to observe it several days in succession to perceive that it had motion . . .
>
> For the rest of this discovery does you much honour; allow me to compliment you for it. I should be very curious, Sir, to learn the details of this discovery, and you will oblige me if you will be so good as to inform me of them.
>
> With equal consideration and respect,
>
> (signed) MESSIER,
> Astronomer to the Navy of France,
> of the Academy of Sciences, Paris[9]

How did Herschel respond to the growing publicity? With some annoyance, one might suspect. He was building a mirror for a new 20 foot long reflector, a project that was never far from his thoughts, and even if a minute could be spared from all the publicity he would get back to it.

On August 31, 1781, Lexell, a celebrated mathematician who had himself discovered a comet, made the stunning announcement of some orbital elements he had determined. He found its perihelion distance to be 16 astronomical units, 16 times the distance between Earth and Sun. This was no comet; William Herschel had discovered the first new planet since ancient times.

Herschel was very self-assured about his discovery. He would have found Uranus sooner or later, he had announced, adding that the method he used was quite well suited to picking up new objects. Later, he would stand toward the top of his larger instrument and sling it one way, then another, observing stars as they went through the field, and making careful drawings of their positions, colors, and whatever else he could find.

Herschel's program involved searching the sky in a methodical manner, recording carefully all that he saw. It was not a comet hunt in the manner of Messier or Pons, but a survey of the sky that Herschel could see with his eye and his telescope.

## 12.3.2 Discovery II: Neptune

> Uranus is a long way out of his course. I mean to find out why. I think I know.
> John Couch Adams, to George Drew, 1841

> That star is not on the map!
> H. d'Arrest, 1846

9 W. Herschel, *Scientific Papers*, p. 86.

The Neptune story actually begins just 14 years after the discovery of Uranus. On May 8, 1795, the French astronomer Joseph Lalande, a superb observer who also spent time popularizing astronomy, plotted a star whose position he noticed two nights later, had changed. Assuming that he had misplotted the object during his first observation, he forgot the matter. Had he checked his field a third time, he would have seen the object move still further. No one would again notice this planet for over half a century.

Soon after its discovery, it became apparent that Uranus's orbit solutions did not fit well with some prediscovery observations that had been made as early as 1690, by John Flamsteed, England's first Astronomer Royal. By 1834, Rev. T. J. Hussey, an amateur astronomer, suggested the presence of 'some disturbing body' whose gravity would be acting on Uranus, in a letter to George Airy, who a year later would become Astronomer Royal. Airy agreed that another planet's gravity might be at work, but since he felt that it would be impossible to find such an object, he discouraged anyone from making a search.

In 1841, John Couch Adams, a 23-year-old Cambridge student, took the problem seriously and after his graduation calculated a position for the new object. He sent his work to James Challis, Cambridge professor of astronomy, who forwarded it to Airy, who expressed interest but did nothing else. When Adams dropped by to visit the Astronomer Royal, he was told that the astronomer could not be disturbed while he was at dinner. Frustrated and unhappy, Adams left his material at the observatory.

Both men behaved somewhat obstinately at this point. Airy did eventually read Adams's calculations and sent the young astronomer a query about them, but Adams felt the question was trivial and never answered it. Meanwhile, he gave his solution to Challis, who had constant access to a refractor of almost 30 cm aperture. Had Challis bothered to look, he should have found the planet. Challis did nothing.

In the meantime, in Paris, Jean Joseph Leverrier had been encouraged by the director of the Paris Observatory, Francois Arago, to investigate the problem, and toward the end of 1845 Leverrier had published a solution. Airy was impressed with Leverrier's work, and when in June, 1846, Leverrier sent him an actual position, Airy noted that it was within one degree of where Adams had predicted. Curiously, Airy never mentioned the parallel work of Adams in his reply to Leverrier. Meanwhile Challis began a cumbersome search based on a wide area of sky around the positions, taking his time and moving very slowly.

Leverrier was unable to get a basic search started at the Paris Observatory, and so he contacted Johann Galle at the Berlin Observatory. At last some luck: Johann Encke, the Observatory Director, authorized a search to begin that very night. Accompanied by Heinrich d'Arrest, a young student, Galle went directly to the predicted position. When he found nothing, d'Arrest produced a chart and carefully they checked the telescopic field star by star. It did not take long before Galle described an eighth magnitude star that d'Arrest did not find on the map, and less than a degree from Leverrier's position! Director Encke was called, and he even measured a small disk of 3.2 arcseconds, only one-tenth arcsecond from Leverrier's size prediction.

Had Challis bothered to reduce his own observations, he would have noticed that he had actually gone over the new planet twice. In any case, when Airy tried later to credit Adams with part of the discovery, the French, led by Arago, were furious. For a while the controversy was spiced up by the French and English press. One happy result was that Adams and Leverrier became firm friends after they met in June, 1847.

### 12.3.3  Discovery III: Pluto

Early in the twentieth century, interest in a ninth planet began to mount. Among several who made calculations, two are prominent: William H. Pickering and Percival Lowell. Pickering had predicted several planets based on a number of factors, including the aphelia (farthest point from the Sun) of some comets.

Lowell's earliest search for a new planet began in 1905 and lasted two years. Using a 5-inch diameter Brashear camera, the search was based on calculations regarding first the comet aphelia, and later on discrepancies in the orbits of Uranus and Neptune. In 1911, Lowell began a second search, this time with a much larger reflector and the use of a new 'blink comparator', an instrument that enables an observer to examine parts of two plates taken of the same region but at different times to check for any change in position or brightness. A year later this search too was stopped. From 1914 to 1916, a third search was mounted using a 22 cm wide-angle camera.

Discouraged by its failure, and exhausted from overwork, Percival Lowell died of a stroke shortly after the third search ended. Reasoning that her late husband had discovered everything worth finding and that his observatory, under the leadership of V. M. Slipher, was not worth supporting, his widow tried unsuccessfully to break his will. The ten-year litigation sapped the observatory's resources and prevented any further search until 1929. Now the work began again by an amateur astronomer from Kansas whom the observatory hired to resume the search, now with a new 33 cm astrographic camera. The new man was Clyde Tombaugh, whose drawings of the planets with his home-made 23 cm reflector impressed the Lowell staff.

Soon after the search began, Tombaugh realized that finding a planet would be hopeless without a plan that paid attention to the retrograde motions of the outer planets. If the search were to concentrate on the parts of the ecliptic where distant planets were moving in retrograde, the speed of 'backward' motion would give away the distance to the planet. A trans-Neptunian planet would thus give itself away instantly.

On February 18, 1930, Clyde Tombaugh was blinking plates he had taken on January 23 and 29. At four o'clock that afternoon, he noticed a 16th magnitude speck appear and disappear on one of the plates, and alternately do the same thing at a different position on the other plate. A check on a third plate he had taken, as well as three plates taken with a smaller camera, also revealed the suspect. The amount of retrograde motion was perfect, and as Tombaugh approached the Director's office, he knew he had it. 'Dr Slipher,' he said, 'I have found your Planet X.'

The announcement to the world was made on March 13, 1930, the 149th anniversary of the discovery of Uranus. One would think that this remarkable story had ended. As time went on, however, it became obvious that Pluto could not have been Lowell's Planet X. It was much too small to have had any appreciable gravitational pull on Uranus or Neptune. Pluto was found as a direct result of Lowell's initiative, but only by coincidence was it near the place of his prediction. It was found because a thorough and careful amateur-turned-professional astronomer conducted the search that led to it.

## 12.4  Observing Uranus

It is incorrect to limit the number of naked eye planets to five, since at its maximum brightness of magnitude 5.5, Uranus can be seen without optical aid by a keen-eyed observer. Through a telescope the planet is easy to discern, mostly because of its strong green color but also because it has a disk – visible under high magnification under nights of good seeing – which averages about 3.5 to 4 arc seconds in diameter.

One of the most interesting visual observations ever made of Uranus was by Stephen James O'Meara, a highly alert observer who was observing Uranus in an attempt to derive its rotation period prior to Voyager II's flyby. In a program that lasted several years and took an inordinate amount of patience, Steve used the 23 cm refractor on the roof of Harvard Observatory, and observed with no filter, and blue and orange filters. After a long time of observing no special or consistent markings, on one twilit 1984 evening he noticed two bright spots that were confirmed by fellow observers Peter Collins and Michael Rudenko. (As Peter Collins would later add, the planet looked a bit like a planetary nebula with a bright double star in the center!)

Figure 12.3. Uranus, drawn on June 9, 1983, at midnight EDT using Harvard College Observatory's 23 cm refractor. Drawing by Stephen James O'Meara.

Uranus is unique in that it is tilted almost on its side, and thus anyone observing a Uranian feature might occasionally see it move around one of the poles. In fact, O'Meara monitored the clouds as they went round in this way, and thereby determined the location of the pole and a rotation period of 16.2 hours. When Voyager II arrived at Uranus in January, 1986, it observed a period of rotation that closely matched that observed by O'Meara.

With a moderate power, giving a magnification of from 150 to 200 times, you may see the disk of this larger of the solar system's two outer giants. As with most planets, a good, long focus telescope of 15–25 cm diameter would be superior to a large but short-focus 'light bucket' of over 30 cm. On March 13, 1981, I celebrated Herschel's discovery 200 years earlier by observing Uranus in a 40 cm telescope, and then in a smaller instrument whose optics were in better condition and whose focal length was longer. Both showed clear disks, but the image offered by the smaller telescope was superior. Planetary work needs good optics, not necessarily big optics.

Typically, Uranus presents a featureless greenish disk that may be quite darkened at the limbs. Sightings of belts and bright spots are rare and require excellent optics, good seeing, sharp eyes, and a lot of observing experience.

Uranus has five moons that are visible telescopically. With their mean visual magnitudes at opposition, they are Miranda (16.5), Ariel (14.4), Umbriel (15.3), Titania (14.0), and Oberon (14.2). With a 20 cm or larger telescope, you may spot Titania and Oberon, the two largest satellites. Incidentally, all but one of Uranus's five major satellites are named for characters from Shakespeare's plays: Ariel is a spirit, and Miranda is the Duke of Milan's daughter, in *The Tempest*; Oberon and Titania are King and Queen of the fairies in *A Midsummer Night's Dream*. Umbriel is the 'melancholy sprite' in Alexander Pope's *The Rape of the Lock*.

## 12.5 Observing Neptune

To confirm spotting this distant blue planet, you should follow it over a few nights. The bluish tinge should be easy to find, but help it along for two nights to be sure of its motion. Similar in some ways to Uranus, Neptune is harder to find because of its much greater solar distance. Its symbol is that of the Roman god of the seas.

Neptune will show virtually no detail through your telescope; in fact it will be a considerable achievement just to discern the disk at all! You should notice its blue color.

Parallel bands, somewhat like those seen on Jupiter, have been seen by astute observers with large telescopes. In fact, some of the early faraway Voyager images showed banding, although the closeup images did not. If they exist at all, they are very elusive; even experienced observers need a lot of time to convince themselves.

Is Neptune's dark spot, discovered by Voyager in 1989, visible from Earth? Although no one has ever reported it, some observers suspect that it should be detectable.

Neptune's two Earth-observed satellites are Triton (magnitude 13.5 at opposition) and Nereid (18.7). Triton should be visible in 20 cm or larger telescopes.

## 12.6   Observing Pluto

Pluto's symbol is an artistic joining of the two letters that stand both for Pluto's name and that of Percival Lowell, who instigated the search.

To follow this object for a few nights to confirm motion is imperative since the 14th magnitude planet can easily be confused with stars that atlases and finder charts do not include. Under a dark sky I have found that with a 15 cm telescope you are pretty well out of luck. A 20 cm reflector will show the distant planet, under a dark sky, if you really know its star field well. With a 25 cm or larger telescope, Pluto should not be too difficult to find.

Pluto's magnitude varies, although its amplitude of no more than a few tenths of a magnitude it not easy to detect visually. However, an advanced project could involve long-term observations of the planet for possible brightness variation. Do not expect to see its moon, Charon – discovered in 1978.

On April 4 and April 5, 1983, I observed Pluto through my 40 cm telescope. Over the two nights the planet inched westward across the high power field. Confirming the faroff planet was an indescribable thrill for me, and it was fun too for my observing partner who wrote in my log that night:

Figure 12.4. Neptune, drawn on July 18, 1983, at midnight EDT using Harvard College Observatory's 38 cm Merz and Mahler refractor, 330×. Drawing by Stephen James O'Meara.

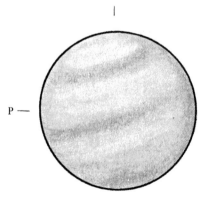

Sure enjoyed looking at the many objects with your 16-inch telescope, especially Pluto, the Hercules Globular cluster, and the Whirlpool Nebula, M51. I could see the spiral structure in M51. Also, M13 was superb! But poor little Pluto was so faint, unimportant looking ... Thank you for this observing session.

Clyde W. Tombaugh

# PART THREE

# Minor Bodies

# 13 Asteroids

There was a time when the minor planets, mostly between Mars and Jupiter, were an inconvenience, a headache. So many of them would appear on photographic plates that precious time would be wasted checking them out; is it a new asteroid, or have I merely rediscovered someone else's? Asteroids were considered a group of orphan rocks in space.

All that is changed now. Asteroid research is one of the most fascinating areas of modern astronomy, partly because of state-of-the-art technology that enables astronomers to use precise light curves and spectral analysis, and even radar tracking, to determine shapes, composition, surface texture, and possible origin.

## 13.1 Historical perspective

Our acquaintance with the asteroids began as a badly needed missing link, a practical answer to a beautifully crafted empirical 'law'. In 1766 Johann Daniel Titius of Wittenburg developed an interesting mathematical formula that appeared to fit the solar distances of each of the known planets.

Titius began with an arbitrary number of 100 'parts' to represent the distance between the Sun and Saturn. Noting that Mercury was 4 of those parts from the Sun, Venus was 7 (or 4+3) Earth 10 (4+6) and Mars 16 (4+12), and that, indeed, the formula worked for Jupiter and Saturn as well, he wondered why doubling the 12 to 24 and adding 4 would not correspond to any known planet at that distance from the Sun.

Meanwhile, Johann Bode, an astronomer and Director of the Berlin Observatory, campaigned for a search for a new planet based on the anomalous result of Titius's formula. So actively did he promote the idea of this 'law', at first without giving due credit to Titius, that the 'law' became known as Bode's law, although it is now known as the Titius–Bode law.

In 1781, when William Herschel discovered Uranus, one of the earliest pronouncements was how the distant planet satisfied the requirements of Bode's 'law', which prescribed a number of 196. The actual number of 191.8 was close enough for the formula to account now for the positions of each known planet. Even though the law's discoverers produced no reason why the 'law' should work, the fact that it apparently did work was reason enough to accept it, and to elevate it to its next step: if the 'law' could explain, could it also predict?

Between Mars and Jupiter there was still an absence of any object, and the conditions of Bode's 'law' were not satisfied there. It was not enough, astronomers thought, to say that planets were at some of the formula's acceptable points; for the 'law' to work at all, there had to be planets at all points. Thus a search was started for the missing planet between Mars and Jupiter by a group of six astronomers in 1800 at the Observatory at Lilienthal. Led by Johann Schroter, and calling themselves the 'Celestial Police', they divided the ecliptic into search regions. Other observers, including Giuseppi Piazzi of the Palermo Observatory in Sicily, Heinrich Olbers, and Karl Harding, were invited to join the search.

Piazzi was ahead of the Police, however. Having undertaken to compile a star catalog, he found an unfamiliar object on January 2, 1801, and the following night noticed some motion. Its orbit seemed to satisfy Bode's 'law', and although its size seemed much too small for a planet, the new object was given the name Ceres.

On March 28, 1802, a second object was found by Olbers. Except for its usually eccentric and inclined orbit, this new body, named Pallas, also seemed to satisfy the Titius–Bode 'law' requirements. (The eccentricity means that the orbit is more of an ellipse than most of the other planets, and the inclination means an orbit inclined at some angle to the plane of the ecliptic.) When a third (Juno) was found on September 2, 1804, by Karl Harding, and finally a fourth (Vesta) by Olbers, satisfaction turned to concern about what could have happened to the mystery planet between Mars and Jupiter. Was it a planet that somehow exploded, or was it pieces of a planet that never got formed in the first place? In any event, since no other objects were found after a lapse of 11 years, the Police surrendered.

For all the excitement about Bode's 'law', it turned out to be little more than an elegant mathematical trick. In any case, the discovery of Neptune showed that distant planet to violate the law; its 300 'parts' are about 88 less than it should be. The formula has been given a lot of historical importance, but had the Celestial Police not bothered to look, Piazzi would have found Ceres anyway at the same time. Olbers and Harding were, of course, members of the group, and their thorough searching firmly established the presence of more than one asteroid.

## 13.2 Naming of asteroids

Generally, the discoverer of an asteroid, or minor planet, gets to name it, but only after its orbit has been determined and it is assigned a number. (The terms 'minor planet' and 'asteroid' are interchangeable. It has evolved into a sort of custom that planetary scientists who study the orbits of these objects refer to them as 'minor planets' and those who study them as physical bodies call them 'asteroids'. A third term, 'planetoid', while correctly descriptive, is not commonly used.)

So many asteroids are known that even giving them provisional designations is complex. The usual procedure involves dividing each year into 24 bimonthly periods, labeled from the letters of the alphabet, although I, a commonly used roman numeral, is omitted. A newly discovered asteroid is given a designation like 1985 QS, where 1985 is the year of discovery, Q represents the 16th bimonthly period (the second half of August) while S represents the 18th discovery during that period. Some of these discoveries turn out to be new observations of previously found objects, while others are never seen again.

The time between discovery and full understanding of the orbit normally lasts through about three oppositions. At that point a number is assigned and then the discoverer names it. Among the exceptions is the case of multiple discoverers, where one might ignore the object after finding it and another spends the time determining the orbit; the latter person might get the honor of the naming. If the discoverer has died, then another appropriate person, perhaps the one who computed the orbit, might get the honor.

A group of asteroids revolves about the Sun at special places, known as 'Lagrangian points' near the orbital plane of Jupiter and the Sun. These asteroids are known as Trojans, and are traditionally given names relating to the Trojan War. The Greeks, with one Trojan 'spy' 624 Hektor, precede Jupiter near the 'L4' point, and the Trojans, along with Greek 'spy' 617 Patroclus, follow Jupiter near 'L5'. In June, 1990, the author, with Henry Holt, discovered the first 'Mars Trojan', 1990 MB.

With about 4000 asteroids numbered, the names span a variety of interests. Although such names are highly uncommon, 2309 Mr Spock was named for its discoverer's cat, named in turn for the television character who, according to the official citation, 'was also imperturbable, logical and had pointed ears'.

Asteroids are named for places (1125 China, 2531 Cambridge), famous astronomers (1134 Kepler, 1501 Baade), planet finders (1996 Adams, 1997 Leverrier, 1604 Tombaugh), composers (1814 Bach, 1815 Beethoven, 1818 Brahms), astronauts (1772 Gagarin, 3352 McAuliffe), writers (2984 Chaucer, 2985 Shakespeare), and rarely for political figures (2807 Karl Marx).

## 13.3  Observing asteroids

As an amateur observer, your greatest satisfaction in asteroid observing may be to add them, one by one, to your list of objects seen, a process just like bird watching. The fainter and more elusive asteroids are as exciting to 'discover' as an uncommon species of bird. Asteroids are easiest to see, and brightest, when they are opposite the position of the Sun (in opposition) and therefore visible in the sky much of a night. Follow your asteroids for several weeks, so you can get a clear idea of where they are going to be in the sky. You may want even to draw a map of stars in the asteroid's vicinity and then plot the interloper's observed path as it moves along.

## 13.4  Kinds of asteroids

For the purpose of our observations, one can say that there are essentially two types of asteroid, in a classification determined by the orbit. The 'mainbelt' objects lie entirely between Mars and Jupiter and include the famous first four objects of Ceres, Pallas, Juno, and Vesta. Other varieties, known as Apollos and Amors, are especially interesting to us and includes such objects as Icarus and Geographos, objects whose perihelia (closest points in their orbits of the Sun) lie near the Earth's orbit. Some of these objects approach Earth; in fact, in 1968, Icarus approached within four million miles of our planet, and in 1989 an asteroid called 1989 FC came to within half a million miles of us. These objects are fun to watch as they hurtle through the sky, their motion sometimes being apparent over a period of a few minutes.

## 13.5  Observing asteroids

The common way to locate an asteroid is first to find its position on one of the charts provided by *Sky and Telescope, Astronomy* magazine or the *Observer's Handbook* of the Royal Astronomical Society of Canada. The charts will show stars with the path of the crossing asteroid drawn in as a line. These sources usually also include the asteroid's magnitude, which you can compare with nearby stars to get an idea of how bright the object should appear. Find the pattern of stars in your telescope's field of view that matches the chart, and then locate the 'star' that appears out of place. 'That's it!' you will probably say excitedly, but prematurely; your search is not yet over until you have confirmed your observation by detecting motion. Draw the field with stars and suspect accurately representing what you see. Wait at least an hour, but preferably several hours, or even a full night, and check the field again to see if your suspect has moved. Your charts should give you an idea of how much daily motion to expect. If the asteroid is in opposition, it will move fairly rapidly. There is a time, however, when the object is near its 'stationary point', and moves very slightly from one night to the next.

Once you have found these bright objects, you should have little difficulty with a few asteroids of fainter magnitude, like Eunomia and Metis. But with fainter objects you will have to spend more time. They are harder to locate, and since they look exactly like stars you will have to make sure that one has moved before you can confirm that you have spotted it. If the asteroid of your search is near at least two stars, motion might be detected quickly. If it is very near one star but far from any others, you can confirm identity only if the star is far from what the asteroid's magnitude should be. Finally, if no stars are nearby, you may need to wait a full night before confirming motion. Once you have some experience, you can try finding fainter asteroids using positions calculated from their orbital elements found in publications such as the *Minor Planet Circulars* and then plotting these positions on a detailed star atlas like *Stellarum*.

## 13.5.1  A life list of asteroids

One of the most interesting things one can do with asteroids is simply to count them, like a list of birds.

Figure 13.1. A form for recording asteroid positions. Courtesy Royal Astronomical Society of Canada, Montreal Centre.

ROYAL ASTRONOMICAL SOCIETY OF CANADA
Montreal Centre

OBSERVATIONS OF ASTEROID................................    OBSERVER.....................................

Use squares below to plot position of asteroid on successive dates in relation to nearby stars, indicating the constellation, giving co-ordinates and identifying brighter stars.

Date................. Time............  Date.................. Time............  Date.................. Time.................
Instrument...........................  Instrument...........................  Instrument...........................
Power.............. ...........  Power...........................  Power...........................
*Predicted Position*
R. A. .............. Decl................  R. A. ..............Decl................  R. A. ................ Decl.............
Remarks:

Table 13.1. *Life list of asteroids*

Observed and recorded by .................................................................................

| Minor planet | | M (blue) | Date | Telescope | Comments |
|---|---|---|---|---|---|
| 1 | Ceres | 7.28 | | | |
| 2 | Pallas | 6.89 | | | |
| 3 | Juno | 7.84 | | | |
| 4 | Vesta | 6.16 | | | |
| 5 | Astraea | 9.97 | | | |
| 6 | Hebe | 8.09 | | | |
| 7 | Iris | 7.54 | | | |
| 8 | Flora | 8.61 | | | |
| 9 | Metis | 9.21 | | | |
| 10 | Hygiea | 9.73 | | | |
| 11 | Parthenope | 9.83 | | | |
| 12 | Victoria | 9.17 | | | |
| 13 | Egeria | 10.02 | | | |
| 14 | Irene | 9.49 | | | |
| 15 | Eunomia | 8.28 | | | |
| 16 | Psyche | 9.85 | | | |
| 17 | Thetis | 10.80 | | | |
| 18 | Melpomene | 8.17 | | | |
| 19 | Fortuna | 9.78 | | | |
| 20 | Massalia | 9.20 | | | |
| 21 | Lutetia | 10.14 | | | |
| 22 | Kalliope | 10.65 | | | |
| 23 | Thalia | 9.37 | | | |
| 24 | Themis | 11.23 | | | |
| 25 | Phocaea | 9.76 | | | |
| 29 | Amphitrite | 9.56 | | | |
| 30 | Urania | 10.31 | | | |
| 39 | Laetitia | 10.16 | | | |
| 40 | Harmonia | 10.10 | | | |
| 41 | Daphne | 9.68 | | | |
| 42 | Isis | 9.63 | | | |
| 43 | Ariadne | 9.91 | | | |
| 44 | Nysa | 9.59 | | | |
| 45 | Eugenia | 10.96 | | | |
| 46 | Hestia | 11.09 | | | |
| 55 | Pandora | 11.03 | | | |
| 65 | Cybele | 11.72 | | | |
| 93 | Minerva | 11.05 | | | |
| 344 | Desiderata | 9.68 | | | |

Table 13.1. *Life list of asteroids* (cont.)

Observed and recorded by .................................................................

| Minor planet | | M (blue) | Date | Telescope | Comments |
|---|---|---|---|---|---|
| 349 | Dembowska | 10.32 | | | |
| 354 | Eleonora | 10.31 | | | |
| 433 | Eros | 7.60 | | | |
| 471 | Papagena | 9.78 | | | |
| 487 | Venetia | 11.93 | | | |
| 511 | Davida | 10.22 | | | |
| 532 | Herculina | 9.04 | | | |
| 654 | Zelinda | 10.07 | | | |
| 679 | Pax | 10.81 | | | |
| 694 | Ekard | 10.84 | | | |
| 1036 | Ganymed | 7.79 | | | |
| 1627 | Ivar | 9.83 | | | |
| 1980 | Tezcatlipoca | 9.93 | | | |

## 13.6 Asteroid occultations

Asteroid work becomes quite exciting and can be of value to professional astronomers if the moving minor planet is about to occult, or pass in front of, a star. Depending on how well the orbit of an object is known, the path of an occultation will be known to a particular degree of certainty; normally any observer within 100 miles of a predicted path has a chance of catching an occultation. And what a sight the occultation can be! Say an 11th magnitude asteroid is about to occult a 7th magnitude star. If you are fortunate enough to be on the path, the star will instantly drop four whole magnitudes in brightness, and stay that way for the few seconds the asteroid takes to cross the point of space where the star is. Because these events rarely last more than 30 seconds, you will not be able to get the two timings with a single stopwatch; you must use a tape recorder to record radio time signals as well as your voice saying 'in' when the star disappears or fades, and 'out' when it reappears.

Using a star atlas or preprinted map in one of the astronomical magazines, locate the star ahead of time. Remember that if you are looking for an occultation you cannot leave the eyepiece even for a few seconds, for it may be in just that short time that the event takes place. Also, it is best to allow 10 or 15 minutes on either side of the predicted time, just in case the asteroid is accompanied by an undiscovered satellite whose presence you would discover as you see a secondary occultation. One night I was observing a star field in hopes of catching an occultation. I saw no event, except that about 20 minutes into my watch, a faint artificial satellite crossed through the field. The following day Derald Nye, an expert occultation observer who had been observing from a site a kilometer away, asked if I had seen the satellite pass by, an event that lasted only a few seconds. My positive response convinced him that I had been watching carefully!

Asteroid occultation observing is exacting work, but even when we fail to see an event it is still fun. Even if you observe and see no event, your negative report does have value in that it shows where the asteroid did not cross and thus can help define its orbit.

## 13.7  Physical observations

As asteroids rotate, they present different sides to us, and their brightnesses change. If the amount of change is less than half a magnitude, you will have difficulty noticing this. Some asteroids change more than this and these can be observed visually for evidence of rotation.

Because the amplitudes are slight, visual estimates of asteroid magnitudes are not usually recommended. This is the province of the observer with a photometer that can measure accurately the amounts of light it collects. Well beyond what a visual observer would normally do, I include the procedure, as part of a story, as an example of how professional observing is done.

### 13.7.1  A photometric study of some asteroids

My first exposure to professional observing happened one cloudy autumn afternoon when I got a call from Clark Chapman of the Planetary Science Institute. They were looking for someone with amateur observing experience to assist in a program of monitoring the light curves of some bright asteroids. Not at all confident in my ability to use a computer, I approached the job as observing assistant with some trepidation.

I needn't have worried. I was working with, besides Clark, three very experienced asteroid observers: Stu Weidenschilling, Rick Greenberg, and Don Davis. Two of us would be at the observatory at any one time. The observing procedure was simple, although the amount of equipment we had to use made it seem quite complicated. We used a telescope known as the 'No. 2 36-inch', one of the smaller reflectors at Kitt Peak. Of all the procedures we had to follow, the one I enjoyed most was the simple act of opening the dome: the two shutters slowly spread apart like the wings of a bird preparing for flight. The opening dome revealed a slowly darkening sky, still bright off to the west.

With the telescope we would begin by finding the comparison star for the first asteroid, and when the star was centered in the photometer, we would then take three five-second 'integrations' during which the photons from the star would be counted. Then we would move off and take three readings on the blank sky background which would later be subtracted from the reading of the star. Next, we would move to the asteroid. Although we had highly accurate positions for our objects, sometimes we would have to wait to determine if what we had thought was the asteroid actually moved in the right direction. We would measure the asteroid and the blank sky background near it in the same manner.

These steps were then repeated on two more comparison stars and asteroids. A cycle of three asteroids observed in this way would take about ten minutes, and we would immediately repeat the cycle so that each asteroid was measured five or six times per hour. We would also read six 'standard stars' whose magnitude values have been accurately determined, at particular altitudes above the horizon. If the sky remained clear we would continue this process, uninterrupted, for up to five hours, and then immediately start again with other asteroids whose observation would take us till dawn. It was hard work, and on clear nights we would have snacks on the job, observing without stopping.

Over a few years of monthly observing runs, we accumulated a vast amount of data that revealed the light curves of several asteroids at different points in their orbits. From all this work we obtained a pretty good understanding of the rotation periods of our asteroids, as well as an idea of their shapes. Some asteroids are thought to be composed of collections of rubble bound together loosely by gravity.

The coldest night I ever spent was during that program. The temperature was not that low, about $-10°C$, but we were in the dome, huddled behind a computer terminal, for over 13 hours. I don't ever recall being so cold. The only sounds we heard were the motions of the telescope and dome as they moved across the sky from one object to another, the beep of the terminal, and our own voices as one observer would tell the other that the asteroid was centered in the photometer and that measuring could begin. As we completed each asteroid observation, a second computer would then plot the portion of the light curve we had already obtained, so even though we saw each object as a faint point of light, we had a clear picture of its rotation as we went through the night. Although I thought I would never warm up after all those hours, when the sky brightened in the east to reveal the star around which both the Earth and these asteroids revolved, I felt that those little planets had let me in on some of their secrets.

# 14 Comets

Old men and comets have been revered for the same reason; their long beards, and pretenses to foretell events.

Jonathan Swift, 'Thoughts on Various Subjects'

## 14.1 Comets, clouds, and variable stars

There are cloudy nights that could clear up later, and then there are cloudy nights where several layers of overcast make things absolutely hopeless for a planned night of hunting for comets. The night of January 4, 1987, was so cloudy that there wasn't a hint of clearing anywhere. The weather map on television showed clouds all the way to the west coast. This would be a night to complete something: for years I had been writing a book about variable stars, and if there were no interruptions, tonight would be the night I'd finish it.

By 3 a.m. I was concentrating almost entirely on variable stars, pausing occasionally to check the sky condition. I recall thinking how comets and variable stars represent such different branches of astronomy; comets are diffuse, fuzzy, and nearby things, and variable stars are distant suns. And yet, visual observers thrive on estimating the changing magnitudes of both. Historically, of all branches of observing, variable star observation and comet discovery are the two areas where amateurs have made the greatest contributions to astronomical thought. In an age where amateur astro-

Figure 14.1. Chris Schur's five-minute exposure of Comet Okazaki–Levy–Rudenko (1989r) was taken through a 20 cm f/1.5 Schmidt camera. Note the straight gas tail.

nomers have larger telescopes and more accessories than ever before, these two fields remain fertile.

By 5 a.m. on the morning of January 5, I typed the last period of the last revised chapter, and the book was finished. Waiting while the computer formatted the material for printing, I checked the sky again. Huge storm clouds were forming in the west, but the east had cleared! With the computer at work on variables, I went to work on comets.

Starting due east, I began searching in a pattern, moving eastward 10 degrees toward the horizon, then one telescope field of view south, then westward about 10 degrees, a field south, and so on. After 15 minutes I went inside to check the computer, and again 15 minutes after that. By now the sky was starting to brighten in the east, and those western clouds were moving in fast. I wondered, as I returned to the eyepiece for a final stretch of hunting, whether the clouds or the dawn would stop this hunt.

It was neither. Two eyepiece fields later I spotted a faint, diffuse patch of light, with an even fainter extension towards the west. Could that be a comet? I would not know that morning, since the brightening sky and clouds rendered the object invisible less than a minute later. Quickly I closed the observatory and went inside to check a detailed star atlas. I did find an object plotted near, but not at, the position of the comet. The night had ended with a completed book, a comet suspect, and rain.

The following night was cloudy as well, and the next day I found the object on my star atlas plotted on the Palomar Observatory Sky Survey. Taken during the 1950s with the large Schmidt telescope at Mount Palomar, this is the best sky atlas of all, the final arbiter of many a suspect. The survey photograph revealed the object as a 15th magnitude galaxy, something my telescope would never show. Moreover, my suspect had a tail!

On the morning of January 7, I found the object a degree south of its first position. Only two hours later, the International Astronomical Union's Central Bureau for Astronomical Telegrams announced the discovery of Comet Levy, 1987a.

This was my second comet, and four more have since come home through my eyepiece. But I'll never forget that strange night where it was cloudy and rained, *Observing Variable Stars* got finished, and somehow in between all that, I found a comet.

## 14.2 Comet observers

Each observing branch offers a special something – the planets with the grace of their motions and the subtlety of their surface detail, variable stars with the suddenness and strangeness of their changes, and the galaxies with their vast distances and massive sizes. But comets are different still. For me, they are special, and of all the types of objects that grace the sky, comets are my favorite.

All observers are different, and all comets are different, but there is but one major requirement – patience – for all observers for all comets. Patience in observing, because comets come in so many different shapes, textures, and apparent sizes, that the object for which we are looking may not be quite what we expect. Patience in drawing, because the appearance and relative brightnesses of a comet's coma, central condensation (if we can see one), and tail can be complicated to draw. Patience in photography, because a comet can move perceptibly among the stars during the course of a time exposure. For seekers of new comets, these vagabonds offer idiosyncratic statistics, cheating some observers out of success after thousands of hours of searching, yet rewarding others after almost no time at all.

Even in a science where every field is known for unexpected surprises, comets are unusual. Their magnitudes, coma diameters, tail lengths, even their orbits offer causes for the unexpected. The person who studies and looks for them is part of a lengthy line of astrologers, soothsayers, poets, observers of nature, and careful scientists.

We are reminded of how little we know about comets almost every time a comet comes our way. Remember Comet Kohoutek, the 'comet of the century' that almost slipped by without being seen? At discovery that comet was much brighter than was typical for comets at its distance, and if it brightened as comets normally do, indeed it would have been the comet of the century. But this was what is known as an 'Oort cloud comet', a comet making its first visit from the huge cloud of comets that surrounds the solar system. Comets on their first visits undergo their initial brightening at greater distances than those that have been here before, and thus Kohoutek appeared a more substantial object than it turned out to be.

In late 1983, Periodic comet Hartley–IRAS (1983v) passed by, brightening by some three magnitudes in a couple of days, and then continuing to defy magnitude predictions for months afterward.

## 14.3  What is a comet?

The wild image of a long, flaming tail and a bright, starlike head is dramatic, but hardly accurate for most comets. It has been some time since we have had a comet of that magnitude, and longer still since we have been treated to a comet of truly gigantic proportions.

Comets are aggregations of ices with interspersed solid rock, like, as Fred Whipple suggested in the 1950s, large 'dirty snowballs'. Typically a comet within about five astronomical units of the Sun (five times the Sun–Earth distance) develops a *coma*, and if the pressure of the 'solar wind' of energy from the Sun on the coma is strong enough, some of the *gas* will push away from the Sun and form a *tail*. Sometimes the *dust* in the coma will form a tail as well; this dust tail is normally curved as its particles begin their own orbits around the Sun.

The best known recent examples of a clear division between gas and dust tails are Comets West and Halley. In March, 1976, Comet West presented especially fine examples of both types of tails. Whether we see a gas, dust, or any tail at all, and how long a tail might be, depends on the actual dimensions of the comet's appendage as well as the angle at which we happen to see the comet.

## 14.4   Families of comets

Comets are members of families if they share, sometime in their history, a capture experience by a major planet. As the most massive planet, Jupiter has captured more comets than any other planet, comets which had, on some earlier passage, approached the giant planet closely enough that its gravity altered the comet orbit so that its period would be shortened drastically. Encke's Comet, with a period of 3.3 years, is probably the best known member of the Jovian family.

## 14.5   Groups of comets

If a number of comets share the same orbit, they are known as a group. The Kreutz group of Sungrazers is a confirmed example. Long ago a comet coming within a few hundred thousand kilometers of the Sun's surface split up into smaller comets, which successively split as they returned, so that now the group has many members. The comet of 1668, comets 1843 I, 1882 II, and 1965 VIII (Ikeya–Seki) are well known examples. A host of smaller comets discovered by the SolWind and Solar Maximum Mission Satellites also belong to this group.

In 1988 Comets Levy (1988e) and Shoemaker–Holt (1988g) were found to share an orbit, and thus became the first instance of independently discovered members of a comet group (or in this case, a pair) other than the Sungrazers. They follow an orbit that brings them back about every 13 000 years.

## 14.6   Observing comets

Would you like to volunteer to keep the neighborhood watch, keeping track of the comets that visit the inner solar system? Many amateurs keep a score card of the number of comets they have seen, as there is some excitement in the competition.

Of the publications that offer positions for newly discovered comets, *Sky and Telescope* and *Astronomy* magazines are the easiest to find. All you then have to do is note the comet's position on a star atlas, and if the comet is visible from your location and time, try to find it. You probably want information faster than the magazines can provide it, however.

The primary source of information is the *Circulars* of the International

Astronomical Union. These cards announce new comets as they are discovered, and the predicted positions and magnitudes. If you can't afford the cost of these, try visiting the astronomy department of your local university frequently; chances are all the latest IAU *Circulars* would be posted on their bulletin board.

Another convenient and economical way is to invest in a weekly telephone call to Skyline, a three-minute telephone recording that provides news of newly found comets and novae, as well as reports of other interesting astronomical news. The current number is published in each issue of *Sky and Telescope*. The recordings are updated every Friday, so that you can take advantage of low weekend rates to reach them. CompuServe, a computer network, also offers comet news through both its Astronomy Forum and *Sky and Telescope*'s news service called Skytel.

The first thing with any comet is to find it. I have absolutely no trouble finding some comets whose comas were condensed, listed at magnitude 11 or even 12. But try as I did, I missed one seventh magnitude interloper strictly because its seven magnitudes happened to be distributed over too wide an area; it had a very low surface brightness. So be careful, and don't be disappointed if a comet succeeds in slipping by you.

Once you have located a comet, record its shape and brightness using numbers and words, pencil and paper, or photographic film.

Figure 14.2. Note the slightly fan-shaped dust tail of Comet Bradfield (1987s) in this 4.5-minute exposure by Dan Ward. He used a 200 mm f/3.5 lens. The comet seems to be 'approaching' the open cluster NGC6633.

### 14.6.1  How to estimate the brightness of a comet

A good way to estimate brightness is by the Sidgwick Method. Devised by a well known British amateur astronomer, this approach involves comparing the brightness of the in-focus comet to those of some out-of-focus stars. Because the comet remains in focus and the stars are defocused, this method is also known as 'in–out'. The method works thus:

(1) Using a star atlas or charts of variable stars, find some stars of known magnitude that seem to be close to the comet's brightness.

(2) Adjust the eyepiece focuser so that the first star you have selected is out of focus and is of the same diameter as the in focus comet.

(3) Memorize the 'average' brightness of the comet's coma, not including the tail, and compare the comet's in-focus image with the star's out-of-focus image. Which is brighter?

Figure 14.3. The *International Comet Quarterly*'s report form for visual observations, reprinted courtesy Daniel W. E. Green.

COMET OBSERVING REPORT FORM. Send completed sheets to INTERNATIONAL COMET QUARTERLY, c/o D. W. E. Green; Smithsonian Astrophysical Observatory; 60 Garden Street; Cambridge, MA 02138, USA. Please convert observing times to decimals of a day in Universal Time (e.g., Aug. 3, 18:00 UT = Aug. 3.75 UT). *Photocopy this sheet for more copies. (This form supercedes all previous forms. 1984 February 1.)*

> *PLEASE PRINT OR TYPE ONLY.*

Name and designation of comet: COMET _____ 19 _____

   Observer _____ Address _____

NOTE: Drawings and additional comments or remarks should be included on separate sheets of paper. To be eligible for publication in the ICQ, columns below marked with an asterisk(*) must be filled in.

| Date (UT)* | M.* M. | Total* Magn. | Ref.* | Instr.* Aperture | Instr.* Type | f/* | Power* | Coma Dia. (') | D. C. | Tail Length (°) | P. A. | Remarks |
|---|---|---|---|---|---|---|---|---|---|---|---|---|
|  |  |  |  |  |  |  |  |  |  |  |  |  |
|  |  |  |  |  |  |  |  |  |  |  |  |  |
|  |  |  |  |  |  |  |  |  |  |  |  |  |
|  |  |  |  |  |  |  |  |  |  |  |  |  |
|  |  |  |  |  |  |  |  |  |  |  |  |  |
|  |  |  |  |  |  |  |  |  |  |  |  |  |
|  |  |  |  |  |  |  |  |  |  |  |  |  |
|  |  |  |  |  |  |  |  |  |  |  |  |  |
|  |  |  |  |  |  |  |  |  |  |  |  |  |
|  |  |  |  |  |  |  |  |  |  |  |  |  |
|  |  |  |  |  |  |  |  |  |  |  |  |  |
|  |  |  |  |  |  |  |  |  |  |  |  |  |
|  |  |  |  |  |  |  |  |  |  |  |  |  |

(4)    Repeat the procedure with other stars until you have compared the comet's brightness with at least two stars, one brighter and one fainter than the comet.

(5)    Estimate the comet's brightness in terms of the two stars. Suppose that one star is 8.5 and the other is 8.9, and the comet is midway between the two in brightness. Then its magnitude is 8.7. If it is slightly closer to the 8.5, then assign it 8.6.

The older Bobrovnikoff Method, also known as 'out–out' because both comet and stars are out-of-focus, is used as well:

(1)    Defocus the telescope until both comet and stars have the same size. As you defocus, you will notice how the stars get larger very quickly while the fuzzy comet changes relatively little in size.

(2)    Compare all the defocused images until you have assigned the comet a magnitude in relation to the stars.

The centers of some comets are highly condensed, and surrounded by outer comas that are much fainter. Since these comets are very hard to estimate by the other methods, you might try a different approach. Developed by Charles Morris, the Morris Method (also known as 'equal out') works as follows:

(1)    Defocus the entire coma just enough that it 'smears' and that its brightness is somewhat uniform.

(2)    Defocus the comparison stars until they are the same *diameter* (not brightness) as that of the comet.

(3)    Compare the brightnesses of the defocused images and estimate the comet's magnitude.

There is one last method. Used by Beyer, and known also as 'way out', it simply proposed that you defocus the comet to an extreme, until it disappeared against the sky background, and then find a comparison star that also would disappear with the same amount of defocusing. Since the brightness of the sky background varies so much, this method probably will not work.

To improve your accuracy using any of these approaches, you should use more than one pair of comparison stars. As with anything else, your ability to make a good estimate will improve with practice.

## 14.7  The coma

A good visual description of a comet includes a description of the coma and tail. Two factors describe its appearance: the diameter and the degree of condensation.

Express the coma diameter in minutes of arc. Draw the coma, complete with field stars, and compare your drawing to a detailed atlas, using the atlas scale to derive the coma size. The degree of condensation simply is a measure of how condensed the coma is. This measurement uses a scale of 0 to 9, where 0 implies a coma with no condensation at all at the center, 3 refers to a coma with gradual increase in central condensation, 6 involves a coma with a bright core, and 9 means a stellar condensation.

By drawing the tail amidst the surrounding field of stars and then identifying the field in a detailed star atlas, you can estimate its length (e.g. 3 arc minutes) and position angle (70 degrees). Position angle begins with 0 at north and continues toward the east (90 degrees).

These data significantly enhance the value of your observation. For example, if an observer records a total visual magnitude (m1)=13.2 for a comet and supplies a coma diameter of 3″ (arc seconds), a second observer reports m1=12.5 and diameter 4″ and a third observer records m1=11.8 with diameter 5″, one can explain the discrepancy by the ability of the later observers to see more comet, because of better telescope or weather conditions. (m1 refers to the magnitude of the entire coma; m2 is a measure of a comet's much fainter nuclear condensation.)

Ultimately, the major advantage of visual magnitude estimates is that human eyes are plentiful, and in many cases the only means of monitoring many comets. A series of visual magnitude estimates has been made for most comets brighter than 12th magnitude, and some of these comets have enough data that light curves over an entire apparition have been prepared. In some cases these light curves represent the only knowledge of a comet's brightness history.

The *International Comet Quarterly*[10] has maintained an archive of visual comet data for some years. Edited by Daniel Green, the *ICQ* has done an excellent job in collecting data from all over the world. This way, anyone doing research on comets can go to a single source for visual data. Because of the care with which the archive has been built, many visual observations can now be used profitably in comet research. If the comet you have been following suddenly brightens by two or three magnitudes, your prompt reporting of the event to the *ICQ* is important.

## 14.8  Comet hunting

The process of searching for comets is a cajoling, hour after hour, to move ever deeper into strange cosmic territory. What will the next field bring? An interesting double star I have not seen before? An edge-on spiral galaxy? The field of one of my favorite variable stars? Or perhaps a comet?

The first confirmed telescopic find was by Gottfried Kirch, from Coberg on November 14, 1680; he had been observing Mars and the Moon at the time. It was not until Charles Messier began finding comets on January 26, 1760, that comet hunting began to develop as a popular sport. Messier developed a reputation for single-minded enthusiasm for his search, finding so many comets that Louis XV would call him 'The Ferret'. After a royal

10 Published by Smithsonian Astrophysical Observatory, 60 Garden Street, Cambridge, Massachusetts 02138.

dinner he went out in the garden to look skyward for comets, an appropriate activity as he had found the comets of 1764 and 1771 with his naked eye. On that night he missed any comet that might have been there: he severely injured himself falling into a wine pit.

We owe Messier an enormous debt. In addition to developing the idea of visual comet seeking, he also created the first catalog of nebulous, or non-stellar, objects. We now have evidence that he recorded 110 objects, providing examples of nebulae, clusters and galaxies all around the sky (see Chapter 18).

On July 11, 1801, Jean Louis Pons discovered his first comet. This comet of 1801 was special for two other reasons: as the first comet of the nineteenth century it earned its first finder an award from Joseph Lalande, and it was Messier's final independent discovery.

Pons called his best comet seeker the 'Grand Chercheur', a short-focus telescope giving a field of view of nearly three degrees. At age 40 he found his first comet during his job as a door-keeper at Marseilles Observatory. After some success as a discoverer of comets he was promoted to a full staff position, and his career total, estimated at 37 independent comet finds (although his name was not attached to all of them), has been unbroken through 1989.

Caroline Herschel had developed a good reputation as her brother William's observing assistant. But on August 1, 1786, she discovered her own comet, an event that surprised the astronomical community and delighted her brother. Eventually she found seven other comets, making her one of the most successful visual comet discoverers.

One of England's greatest amateur astronomers, William F. Denning, began his astronomical career by starting a 'Society for Young Amateur Observers' in 1869. Remaining a bachelor throughout his life (supposedly to allow him more time for observing), Denning excelled in both meteor and comet observing. His five comets include a periodic comet which was lost and found again recently by the Japanese observer Fujikawa. Now known as Periodic Comet Denning–Fujikawa, the comet is a member of Jupiter's family.

E. E. Barnard (1857–1923) discovered 16 comets, the last of which was also the first comet to be found on a photographic plate taken to survey a star field. An indefatigable observer, Barnard took advantage of a comet award offered by H. H. Warner; comets would come along just as a payment would be due for his new house. 'The faithful comet,' he wrote,

> like the goose that laid the golden egg, conveniently timed its appearance to coincide with the advent of these dreaded notes. And thus it finally came about that the house was built entirely of comets. This fact goes to prove the great error of those scientific men who figure out that a comet is but a flimsy affair after all, infinitely more rare than the breath of the morning air, for here was a strong compact house, albeit a small one, built entirely out of them.[11]

11 Mary Proctor, *The Romance of Comets*, p. 27. New York: Harper and Brothers, 1926.

The South African observer William Reid did not report one of his discoveries. After telling his friends that he had found the comet, he then heard that it had later been picked up by someone else. Even though he had seen it first, he declined to report it officially, and deny credit to the other hunter.

William R. Brooks, the village photographer of Phelps, New York, and one of the most prolific American discoverers, became ill and died after an intense observing period with Periodic Comet Pons–Winnecke 1921 III. Incidentally, on its next return in 1927, it was the first comet observed by an amateur astronomer, Clyde Tombaugh, who three years later would discover Pluto.

Leslie Peltier began his comet hunt with a 15 cm refractor that was lent to him from Princeton University, and when he learned that his new instrument had an honored past, with three finds by Zaccheus Daniel, he decided to continue its tradition. Three years later, on November 13, 1925, he discovered his first comet, and over the next three decades he found 11 more, one of which became bright. Peltier's autobiography, *Starlight Nights*, is a symphony to the joy of pursuit of comets, and reading it may give you a feel for the project you are about to undertake (see Chapter 24 for more details).

On September 18, 1965, Kaoru Ikeya discovered an eighth magnitude patch of haze in Hydra, and Tstomu Seki discovered the same comet independently only one hour later. Shortly after the discovery of Comet Ikeya–Seki was announced, an orbit was calculated that projected the object to approach within three hundred thousand miles of the solar surface, a hairpin turn that would result in the comet's becoming extremely bright. The cometary show was the most spectacular in more than half a century.

From time to time award programs honor discoverers of comets. In 1831 King Frederick VI of Denmark introduced a Gold Medal, which continued until Christian VIII in 1848, and later supported by the Vienna Academy of Sciences, until 1880. H. H. Warner started giving $200 per comet towards the end of the nineteenth century, and the Donohoe Medal of the Astronomical Society of the Pacific lasted until the 1950s. During the 1980s, Roger Tuthill sponsored an award program.

## 14.9  Procedures for hunting

If you wish to have a continuing supply of fresh comets, you need keep in mind only three things:

(1) *Keep the covers off your telescope.* A telescope in a closet is not going to find any comets. Far more important than the kind of instrument you use for comet hunting is the recognition that you need to spend many, many hours with that telescope in your search.

(2) *Know what a comet looks like.* Observe as many comets as you can, so that when you find your own you will recognize it as a comet and not a

galaxy, or the ghost of a bright star. Too many would-be discoverers ignore this maxim, thinking that with absolutely no experience observing comets they will still recognize one. After some years, they give up the ghost, so to speak, or maybe report that ghost as a comet!

(3) *Know what a comet does not look like.* You need to become a deep-sky observer, studying all the different types of objects and seeing how they differ from comets. Most globular clusters have a mottled appearance; most galaxies have a sharp boundary, while that of a comet is very gradual.

### 14.9.1  Sun vicinity

Here is a simple search project. Place a street lamp between you and the Sun, and scan the area within 20 degrees of it. No telescope is required. Rarely, comets have been discovered this way, most notably the bright comet 1910a, which had approached the Sun from the opposite direction from us. After perihelion it was noticed in daylight by some South African railroad workers. In 1967 someone reported to me that he had found such a comet; it probably was a long cirrus cloud.

### 14.9.2  Twilight horizon

Searching for bright comets that are visible in the evening sky as it darkens, or the morning sky at dawn, with binoculars, can be successful, as the story about Walter Scott Houston and Comet Mrkos shows (see Chapter 2). Search the horizon about a half hour after the Sun has gone down or before sunrise.

### 14.9.3  A group search program

This program is best set up by a club, or at least a few dedicated observers working together. In one version, sponsored by a group in Montreal, Canada, the entire visible sky was divided into 428 areas, each to be checked down to sixth magnitude on a clear night for comets, or, for some areas, novae. Frankly, the chances for a comet discovery in this way are very low unless the entire sky is covered, especially the areas in the early evening sky in the west, or low in the east in the pre-dawn sky.

## 14.10 Hunting with a telescope

If you really are serious about hunting for comets, you need to use a telescope. Hunting with a telescope is far more demanding on your time and energy than the other ways of searching, but it also could be the most rewarding.

For this activity you should have at least a 15 cm telescope that is capable of a field of about one degree, a dark country sky with no moon, and an inordinate amount of patience. If you hunt with the expectation that you will soon find something, you will be disappointed quickly. On the average, observers hunt about 400 hours for the first comet and 200 for each additional one, although Mori found his second comet about 70 minutes after his first, and Don Machholz spent over 1700 hours for each of his first two comets.

The better you know the sky, the more fun the hunt. You should not have to stop to check the more obvious Messier objects; if you have to stop too often the work can become tedious. For this reason I recommend that you do not begin a serious comet hunt until you have located and described every one of the Messier objects.

### 14.10.1 Search procedures

Comet hunting is simple, involving either horizontal sweeps, or searching in a vertical pattern up and then down and then up again, or moving the telescope along arcs of right ascension or declination. Whether you choose one pattern or the other is a matter of personal preference. Search carefully each field of view, moving your telescope in one direction across some 40 degrees of sky. Then swing back to the starting point and hunt the area west or east. I spend between one and 15 seconds on each field, depending on how densely populated it is. If you hunt areas of the sky with many galaxies, like the Coma-Virgo regions, be prepared to check the identities of each. If all the objects in the field are star-like, then move on to the next field, and the next. Sooner or later a 'fuzzy' object will sail by. Don't panic; just check it according to a good atlas; you most likely have found a galaxy, cluster, or nebula. Even if your atlas shows nothing, you must check other sources, like a better atlas or catalog of objects for identification. And finally, the litmus test: is the fuzzy patch moving?

Although comet hunting can be done with almost any telescope, instruments with short focal lengths are better for two reasons: they enable you to see more of the sky in one look, shortening the search time; and their wide fields increase the contrast between nebulous patches and sky background, increasing the chance that you will spot a faint comet. Conventional wisdom has always recommended altazimuth over equatorial mounts, since the simpler variety gives you more freedom to search comfortably down to the horizon. The equatorial mount, so useful in tracking objects as they cross the sky, becomes an encumbrance as you search for the easiest way to move up and down the sky in leisurely sweeps.

Rolf Meier of Ottawa, Canada, does not accept this 'conventional wisdom'. Using a 40 cm telescope, mounted equatorially so that he can sweep along right ascension or declination arcs, he found four comets within 200 hours of searching.

## 14.11 Appropriate times

Although comet hunting is a leisurely pursuit, it does require some judgment in choosing the best times for sweeping. If the Moon is in the sky, chances for a discovery become so infinitesimal that such time is better spent in doing other things. Also, you should understand that the sky around the meridian at midnight is the territory of the big photographic telescopes. Comets found in the middle of the night, far from the Sun, tend to be fainter than 12th magnitude, and are found on the photographic plates taken by professional astronomers with large Schmidt telescopes or other instruments capable of showing wide areas of the sky down to faint magnitudes.

These are guidelines meant, it might appear, to be broken. Yanaka discovered his first comet, 1988r, with a bright gibbous moon in the sky. William Sorrells found Comet Sorrells in the middle of the region where professional astronomers observe, and George Alcock found Comet IRAS–Araki–Alcock 1983e as a bright object with binoculars where only faint comets are usually found.

Although it is rare for an amateur discovery to take place against such formidable competition, one reason is that few amateurs even dare to compete. I am not recommending against midnight searches, especially with the big 40+ cm reflectors that are now so widely used by amateurs. Just know your competition and do not expect instant success.

The best time for comet sweeping is in the morning before dawn, during the two days before and the five after new moon. Here is fresh territory, not well covered by professionals, where new comets can make their debuts as they come from behind the Sun. Four times as many comets are discovered in this part of the sky as in the corresponding dusk regions.

Try to observe as much of the sky each month as possible, beginning with the regions within 90 degrees of the Sun; since comets are brightest when near the Sun, they are much more likely to be found by amateurs searching this region. Overlapping and repeating searches during a month is a good idea, especially in the sky nearest the Sun, for you may easily have missed a faint comet.

In the evening sky, begin as soon after full moon as possible, and in the morning sky, begin a few nights before new moon.

## 14.12 Discovery

You think you have found a comet. What do you do?

Look carefully through the eyepiece, and pay attention to what the object tells you about itself. Is it symmetrical? If yes, chances are that you are looking at a galaxy or cluster. Is it condensed or diffuse? Most, but not all, comets appear diffuse. If it has a mottled, circular appearance, it might be a globular cluster. Does it have a tail? (It could still be something like Hubble's variable nebula, which resembles a comet with a short tail.) Is there a bright star or planet nearby, and does the object move among the stars as you move the telescope? If it does, it is likely a ghost of the bright star.

Draw four sketches, showing its position among some naked eye stars, its position in the finder, its position in the low power, and its exact position in a high power eyepiece, in relation to nearby eyepiece field stars. If the comet is setting, twilight is advancing, or clouds are approaching, you may not have the time to do all this.

Check the position in a star atlas or catalog. I find that *Skalnate Pleso*, or the new Tirion *Sky Atlas 2000.0*, show over 90% of the suspects I find in my 40 cm telescope. *Uranometria 2000.0* is even more complete. But say the object is not shown on any of these sources, and its diffuse appearance still makes you suspicious.

Check the object again after at least a quarter of an hour; in any case as long as you can, depending on the constraints of twilight, clouds, or the object's setting time. When you next look at it, you should notice if it has moved. If there has been no motion by the time dawn or object setting has occurred, I advise strongly against sending a telegram. It is a fact that all comets do show motion eventually. You should wait to confirm motion, even if that process means you must wait for another day.

*Don't cry wolf: confirm first.* If the Central Bureau reads about your discovery of the Andromeda Galaxy or the globular cluster in Hercules, it will not be inclined to take you seriously the next time. The philosophy of sending your announcement telegram, as some people have done, just to make sure you are the first, is hardly a scientific procedure and often can embarrass you.

Have you found an already known comet? Where are the known comets? Using the *Observer's Handbooks* of the British Astronomical Association or the Royal Astronomical Society of Canada, current magazines, or a collection of recent IAU *Circulars*, check even the positions of comets that should be too faint for your telescope; one of them may have surged in brightness. If this is the case, you may have discovered something remarkable about this known comet, and should report the observation.

If possible, have your sighting verified by a reputable observer. Tell that person the position and suspected nature, and direction of motion, of the object, and ask him or her to confirm it when possible.

If you are certain the object has moved, and no previously known objects are reported to be in the area, and your observation has been confirmed, you must prepare to notify the Central Bureau of Astronomical Telegrams in Cambridge, Massachusetts.[12]

12  You can reach the Bureau by telegram or telex: (TWX) 710-320-6842 ASTROGRAM CAM. Definitely follow up with a letter addressed to the Central Bureau at 60 Garden Street, Cambridge, Mass., 02138, USA.

They need the following information:

(1)  Your name and suspected nature of object.

(2)  Date and time of observation. Although it is customary to convert to tenths of a day in Universal time (23.5 means 12:00 on the 23rd), just make sure you have the date and time reported accurately.

(3)  Position in 1950 co-ordinates: right ascension and declination. If you prefer to use 2000 co-ordinates, don't forget to state that!

(4)  Direction and rate of motion. This is important, since it may be some time before a clear sky somewhere permits someone to confirm your finding. Either express this motion in arc minutes per hour, or better, simply supply two good positions.

(5)  Magnitude and physical description.

A fictitious example:

> Karen Silverman reports her discovery of a comet, as follows:
>
>    1989 August 23.50 RA (1950) 23h 47m.3 Dec +34.1 degrees.
>             24.39              23h 45m.2       +33.3
>
> Magnitude 8.7; Coma diameter 1 arcmin. Diffuse, slight condensation, tail 3 arcminutes in P.A. 70 degrees.
>
> Karen Silverman, 120 Comet Drive, Tucson, Arizona.

If your discovery is confirmed by the IAU Central Telegram Bureau, then the comet will be named for you, unless others have also found it. Traditionally, comets are allowed a maximum of three names that represent the first people to have seen and reported it *before* it is announced, and therefore you are in competition with other amateurs, professionals using photographic plates and CCD imaging systems, and even orbiting satellites. During its short lifetime the Infrared Astronomical Satellite discovered several comets, helping to make 1983 a year with a big harvest of comets. In 1987, 35 comets were found; about half were discoveries while the others were returning periodic comets.

## 14.13 The naming of comets

In 1939, Faber and Faber published a delightful book of poems by T. S. Eliot, one of the twentieth century's greatest English poets. Called *Old Possum's Book of Practical Cats*, this collection includes a poem called 'The Naming of Cats', which proposes that all cats have three names.

Comets also have three names. After discovery, a comet is assigned the name of its discoverer or discoverers, and also a designation based on the year and order of discovery. Thus, Periodic Comet Halley, 1982i, is the ninth comet to be found (it was recovered, not discovered) in 1982. After all the comets of a given year are listed, they are named again in order of the time of perihelion passage. Comet 1982i is also 1986 III, the third comet to round the Sun in 1986.

Figure 14.4. Comet West. Five-minute exposure by the author.

If comets are substituted for cats, would T. S. Eliot's poem have read a bit like this?

The naming of Comets is a difficult matter,
   It isn't just one of your holiday games;
You may think at first I'm mad as a hatter
When I tell you, a comet has THREE DIFFERENT NAMES.
First of all, there's the name that the family use daily,
   Such as Whipple, Wilk–Peltier, Wirtanen or Wolf,
Such as Hubble or Humason, Honda, P/Halley –
   All of them sensible everyday names.
There are fancier names if you think they sound sweeter,
   Some for the gentlemen, some for the dames:
Such as Grigg–Skjellerup, de Kock–Paraskevopoulos,
   Schwassmann–Wachmann, Herschel–Rigollet, Tsuchinshan 1,
Churyumov–Solodovnikov, Bappu–Bok–Newkirk –
But all of them sensible everyday names.
But I tell you, a comet needs a name that's particular,
   A name that's peculiar, and more dignified,
Else how can he keep up his tail antisolar,
   Or spread out emissions, or cherish his pride?
Of names of this kind, I can give you a score.
   1910a, '84u, '86b and such,
Or '65f, '66b, '83d – there's more –
   Names that never belong to more than one comet.
But above and beyond there's still one name left over,
   And that is the name that at first you can't guess;
The name that no human research can discover –
   Until long after the comet's come and it's gone,
Like nineteen hundred and fifty-nine X –
   But the COMET HIMSELF KNOWS, and won't now confess.
When you notice a comet in profound meditation,
   The reason, I tell you, is always the same:
His mind is engaged in rapt contemplation
   Of the thought, of the thought, of the thought of his name:
     His rotational, orbital,
     Coma-morphological
Deep and inscrutable, singular Name.

# PART FOUR

# Deep sky

## 15 Double stars

'What does a star *look like* in a telescope?' Each time a child puts this question to me, I would show a single star like Arcturus or Vega, and I would get the same puzzled stare. 'All this power, and the star still looks just as it did without a telescope!'

I do not have that problem any more. One evening, after a brief explanation of starlight, and of why a telescope will not show a star, at its great distance, as more than a 'point source' of light, I then showed the children how a telescope can show *more*. 'See Albireo, that nondescript star up there in the middle of the summer triangle?' After a few minutes all the children had found it. Look what a telescope does to that!

This is the magic of double stars. A good proportion of the stars in our galaxy are multiple systems, ranging from two to half a dozen members. Their separation, their contrasting colors, their predominance in almost any star field they grace, offer an ideal way to introduce a beginner to the beauty of what lies beyond the solar system.

## 15.1 Mizar

Because it is in a famous constellation, the most famous double star is Mizar, the middle star of the Big Dipper's handle. It will probably be the first double you look at. Try looking toward the middle star of the Dipper's handle. At first you will be surprised how easy Mizar is to separate, for its apparent companion lies just a short distance off to the northwest. When I first saw this little gem I was pleased to have identified my first double. Actually, I had not. The companion star I had seen, and which you will see, really has nothing to do with Mizar at all; the faint 'companion' star, called Alcor, has no physical connection to bright Mizar. It seems to be nearby only because it is roughly on the same line of sight as its brighter neighbor. Unfortunately we use the term 'double' too loosely, to describe any close pair of stars, whether the arrangement is by coincidence or real. Doubles whose appearances are due to chance alignments are known as optical doubles, while truly related stars, gravitationally bound systems whose members orbit one another, are known as binary systems. Because Mizar and Alcor look like a double star, we call them an 'optical' double: they appear that way only by celestial coincidence.

Point your telescope toward Mizar, which is always available for most northern hemisphere viewers. You will actually see three stars, bright Mizar, fainter Alcor, and a third star fainter than both. If your telescope's optics are good, and if the sky is steady, you should see Mizar itself as two stars sharply separated.

As you study Mizar through your telescope, think a bit about its importance in double star history. You can almost trace the study of doubles through a historical study of Mizar. The first star to be recognized as double through a telescope, in 1650 Joannes Baptista Riccioli observed Mizar and Alcor and saw that Mizar had a companion. Mizar was the first double star to be photographed (by George Bond at Harvard in 1857), and the first double to be studied through a spectroscope (by E. C. Pickering, also at Harvard, in 1889). Tonight, Mizar will be the first double to be observed by you.

In northern hemisphere spring, summer, and autumn, your next target should be Beta Cygni, also called Albireo. One of the finest sights of the entire sky, Albireo is a real showpiece. Its two components shine with different brightnesses and colors, the brighter one a soft yellow, the fainter one a light blue. The contrasting colors and stateliness of two stars placed next to each other is breathtaking. Recent findings have shown that this beautiful pair is an optical double and not a true, gravitationally bound, binary system. Although for me, this news was tantamount to learning that there is no Santa Claus, Albireo is still one of the best of a huge number of stunning stars whose different separations, colors, and relative brightnesses offer hour after hour of enjoyable sightseeing.

Another marvellous system is visible from both hemispheres. In northern hemisphere winter, southern hemisphere summer, observe the quadruple star Theta Orionis, at the center of the Orion Nebula M42.

## 15.2  Historical notes

The first star ever to be identified as double is much further south, in Sagittarius. This is a little-known pair called 'Nu 1' and 'Nu 2' Sagittarii, a pair the astronomer Claudius Ptolemy described as double around 140 AD. The study of doubles really gained momentum at the start of the nineteenth century when, in 1803, William Herschel discovered that the bright components of Castor and of Gamma Virginis had changed their positions relative to each other, and that this change was due to the orbital motion of the stars around one another.

With the start of his sky survey around 1776, William Herschel became intensely interested in listing any new double he could find, and his early nineteenth century publications were the result of years of careful observations in which he would measure the parallax of double stars by comparing the position of the brighter star of a double with the fainter. In 1824, Wilhelm Struve began a program to measure the positions and separations of what would turn out to be over 3100 pairs. His son Otto continued this work.

An amateur astronomer with a 13 cm telescope in his back yard and a passion for double stars became one of the most intriguing figures in the history of this field of study. Sherbourne Wesley Burnham began by looking for new pairs of doubles that had not been previously catalogued. As his observing experience increased over almost 50 years, he found more and more examples of double stars, some already known and needing better measurements, and about 1300 never before noticed. His catalog of doubles, published in 1906, listed 13 665 pairs of his carefully observed doubles. The thoroughness of Burnham's work remains an example of what amateurs are capable of. The substance of this work, the careful measurement of distances and position angles of doubles, is still appropriate today since these positions change as the stars orbit each other.

## 15.3  Nature of doubles

What exactly is a double star? The question is worth some consideration, since over half the stars in our galaxy are double or multiple (three members or more) systems. Castor, the brightest star in Gemini, is a family of six stars, two of which can be seen using a moderate telescope. Each of these two has a companion, and there is another faint pair nearby that is a part of the whole system. Over the next few dozen years, the two bright stars will move further apart.

Once a star is determined to be a real system, astronomers then try to see if their member stars are getting closer or farther apart as they orbit each other. Double stars are meant for a patient observer. Unless you plan to observe components of double stars over a period of decades, or for some pairs, centuries, you won't find much change in their positions relative to one another. Their orbits are large and their revolution periods are very long. One night in our Montreal Centre's observatory the person in charge had just explained about orbits and periods, when one of the newer members, evidently unaware of the star's 600-year period, said 'Oh, yes, I can see them revolving!'

If we choose to classify true binaries according to how we can detect them, there are four types of double systems. The *visual* binary type has components that are separated by enough space that we can see them as a double through a telescope. An *astrometric* system has been identified by the accurate measurement of a component's position over many years. If we decide, after measuring these positions over time, that it seems to move about an unseen point that is not its center, then we can assume the presence of a companion. Other systems have been detected because a spectroscope has analyzed their light to reveal two separate stars; we call these *spectroscopic* binaries. When we detect a new system by the eclipse of one member by another, we have another type calling an *eclipsing* binary.

Algol, the 'Demon star' of Perseus, is an eclipsing binary with an intriguing story, which began in 1782 when John Goodricke first observed its periodic fluctuations and even suggested that the cause of variation may be due to a second star, a companion, passing in front of the brighter one. In 1889, H. C. Vogel of Potsdam confirmed Goodricke's work when his spectroscope revealed that the star was double. Because one member of the binary, or double, system passes in front of the other, causing an eclipse, we call Algol an eclipsing binary.

## 15.4   Observing double stars

In a time when most observing is better accomplished using photography or CCDs, visual observing is still important with double stars. Through visual observations, carefully obtained over many years, we can determine the orbital motion of binary star systems and deduce from that the masses of each of the member stars. There is another factor. When you photograph a faint double star, you need to expose your film for at least a few minutes in order to gather sufficient light so that the stars actually appear on your film. Unfortunately these long exposures also give the currents and eddies of Earth's atmosphere a chance to affect the images. We have already discussed this phenomenon of *seeing* (see Chapter 9). Thus, a star's image will appear not as a point of light but as a small disk, with its neighboring companion either unresolved or completely invisible in its glare. The human eye does not necessarily need long periods to integrate, and can catch a second or two of excellent seeing just long enough to push the telescope to its limit and split a close double. In the basic task of splitting double stars, visual observing wins over photography.

Separating close pairs of doubles will teach you something about how well your telescope will resolve. Before you encounter a double, you will need to know how bright each member is, how far apart they are, and the position angle in which they present themselves to us. Their brightnesses are expressed in magnitudes, so that two stars of magnitudes 5.5 and 5.6 are almost equally bright, while a pair 5.7 and 7.6 are definitely not; the 5.7 star being much brighter than the 7.6 one. We express the separation in terms of seconds of arc, where a second is one sixtieth of a minute and a minute is one sixtieth of a degree and 360 degrees cover a 'great circle' in the sky.

Triple and quadruple star systems can add zest to any evening's celestial sightseeing party. Epsilon Lyrae is a double that can be resolved on a good night with the unaided eye. But did you know that a 15 cm telescope will reveal a secret faint companion for each of the main components? Be sure to visit some night with my favorite triple Sigma 2816 in Cepheus. In the same field of a low-power eyepiece should be Sigma 2819, a double. A 15 cm telescope should split both sets easily. For fun, try Gamma Leonis, Gamma Andromedae, Gamma Virginis . . . it seems that the gammas have a double monopoly!

### 15.4.1 Recording your observations

When you observe a double, record the following information:

*Date*: Record the date and time of your observation as follows: 1987 08 17 03 17. This means that you observed the star on August 17, 1987, at 03h 17m Universal time. (Do not forget to convert your local date and time into UT by adding or subtracting the correct number of hours. If your observation is at 17 minutes past 8 p.m. Mountain Time on August 16, add seven hours to get 03h 17m, but now August 17. If you do not wish to convert, specify the time zone you use.)

*Published information*: For your own reference, write down the published *magnitudes*, *position angle*, and *separation* of the two components.

*Weather*: Your ability to separate a double star will depend partially on the weather conditions, particularly the seeing. To record these conditions at the time of observation is very useful; you will learn under what conditions a star can be split.

*Observed separation*: Finally, you need to record the important fact of whether you were able to separate the star, and what power you needed to make the split. If you were not able to divide the star, record that too. Did the star look completely single, for instance, or could you occasionally see that the star looked elongated, indicating the presence of a companion?

### 15.4.2 Doubles as optical tests

Even if, for some reason, you have not the slightest concern for these star systems, you can use them as tests for the quality of your telescope. Doubles have long been considered a good test for this, for a mirror or lens is judged partly on how close a double it can separate. For this purpose you should find doubles whose components are of approximately equal magnitude, and whose separation approaches the limit of resolution set for a telescope. Should you try to split stars where one is much brighter than the other, the brighter one may obliterate the fainter. A clean separation of the star at your telescope's limit gives you the satisfaction of knowing that your telescope has excellent optics. You have already started with Mizar and know how its two components, of almost equal brightness, complement each other. Also, choose stars not far from the limit of your telescope's capability. Why this strange advice? If you choose stars that are too bright, their light will swamp the tiny space between them, making splitting more difficult.

With what separation of double should you begin? Nothing too challenging at first: test your telescope's limits gradually by starting with something really easy, like Albireo, which has a 30 arc second separation. Table 15.2 suggests some pairs to try. A double that is closer together, and then one still closer, will challenge your telescope until you find one that is so close together that its members appear jointly as a single elongated star image. This is not necessarily the limit of your telescope, only something that approaches the limit on a night that may not have the best seeing. There is a theoretical limit, an equation devised by the nineteenth century 'eagle-eyed' observer William Butler Dawes to define the limits for a good telescope using stars equally bright at about sixth magnitude. A 6 cm refractor of the type that department stores sell should resolve at least to 2 arc seconds, or even a bit better. A 7.5 cm telescope will resolve stars down to a limit of 1.5 arc seconds, and a 10 cm working to its theoretical efficiency will resolve stars separated by just over one second. If your telescope does not perform to this limit, don't worry yet: the cause may have nothing to do with the telescope. Try again later when the seeing may be more steady, or the star higher above the horizon. If the sky is unsteady, the burbling and shaking of the image will seriously impair what your telescope will show. Choosing a pair of unequal brightness, or a pair that is faint, will also make separation much more difficult.

### 15.4.3  The Tombaugh–Smith seeing scale

In addition to seeing details on a planet's surface, seeing quality sharply influences a telescope's ability to separate the components of double stars. In Chapter 9, I discussed a rudimentary way of measuring the effects of seeing. Now that we have more experience, why not consider a more accurate way of measuring seeing? Developed by Clyde Tombaugh and Brad Smith and published in the July 1958 issue of *Sky and Telescope*,[1] this seeing scale (Table 15.1) describes the effect of the atmosphere's conditions on a telescope's ability to resolve a series of double stars located close to the north celestial pole, and thus always visible for northern hemisphere observers. Although many observers are happy to report seeing on a subjective scale, recording seeing by testing with double stars is more accurate, and also invites you to become more interested in observing double stars.

If you are interested in estimating seeing to this accuracy, Table 15.2 includes these north circumpolar stars. If you look at the double star 20 Draconis, for example, whose components are separated by just 0.6 arc second, and notice that the two stars seem to touch each other, then the seeing, according to Table 15.1, is between 5 and 6.

---

1  C. Tombaugh and B. Smith, 'A seeing scale for visual observers', *Sky and Telescope* 9 (July, 1958), p. 449.

Table 15.1. *The Tombaugh–Smith seeing scale*

| Quality | Image diameter (arc seconds) |
|---------|------------------------------|
| −4 | 50 |
| −3 | 32 |
| −2 | 20 |
| −1 | 12.6 |
| 0 | 7.9 |
| 1 | 5.0 |
| 2 | 3.2 |
| 3 | 2.0 |
| 4 | 1.3 |
| 5 | 0.79 |
| 6 | 0.50 |
| 7 | 0.32 |
| 8 | 0.20 |
| 9 | 0.13 |

Table 15.2. *Double stars for the Tombaugh–Smith seeing scale*

| Star | Magnitudes | | Position (angle degrees) | Separation (arc seconds) |
|------|-----------|-----|--------------------------|--------------------------|
| Psi Draconis | 4.0 | 5.2 | 15 | 30.6 |
| Sigma 485 | 6.1 | 6.2 | 304 | 18.3 |
| Xi Cephei | 4.7 | 6.5 | 279 | 7.3 |
| Sigma 2948 | 7.0 | 8.7 | 5 | 2.8 |
| Sigma 2950 | 6.0 | 7.2 | 302 | 2.3 |
| Sigma 185 | 7.0 | 8.5 | 20 | 1.3 |
| Sigma 2054 | 5.7 | 6.9 | 355 | 1.2 |
| Sigma 460 | 5.2 | 6.1 | 65 | 0.9 |
| 20 Draconis | 6.5 | 7.1 | 78 | 0.6 |
| Omicron Sigma 52 | 6.4 | 7.0 | 94 | 0.5 |

Table 15.2, which lists suggested Tombaugh–Smith stars, is useful not only for seeing check, but also to evaluate your telescope's own ability at splitting stars of different separations. The position angles are included to assist with the more difficult pairs. Stars marked 'Sigma' are from the *Wilhelm Struve Catalog*; those labeled 'Omicron Sigma' are from the later listing by Otto Struve.

## 15.5 Advanced work

Would you like to measure precisely the position angles and separations of some binaries that change relatively rapidly? The fourth and fifth magnitude components of Xi Ursae Majoris, for example, revolve about each other in 'only' 60 years, and the third and sixth magnitude members of Zeta Herculis, though difficult to separate, orbit in a relatively rapid 35 years.

There are at least two ways to observe these orbits over a long period. One involves a detailed star atlas, and the other uses a device known as a *filar micrometer*. The idea is to build up a series of measurements that you would repeat several years later to see if any changes have occurred.

A star atlas that shows enough faint stars will allow you to determine the orientation of our field of view. *Uranometria 2000.0* might be sufficient, although a more detailed atlas like *Stellarum* is better for telescopes of 30 cm or larger. By plotting the field and comparing the double star with the surrounding stars, an approximation of the position angle can be estimated.

A filar micrometer attached to a telescope's eyepiece will measure accurately the angular separation of a double, with a view to recording changes over a long period of time. Results of such careful work can have scientific value.

A micrometer is not a complicated instrument. It consists of two very thin wires, one fixed, the other movable, arranged at the eyepiece in such a way that its user can determine the angular distance between the wires and their orientation with respect to north and south. The key to successful micrometer usage is to attach it to the best telescope, preferably a refractor of long focal length, that you can find. First you should calibrate the system by finding its north; then you determine how many seconds of arc a full turn of the micrometer screw will yield.

Probably the hardest part of the observation is this calibration. Using a low power and with the drive off, let a star trail along the fixed wire. By rotating the box you should eventually see the star travel right along the wire without deviating away from it. Moving 90 degrees from that angle will give the north point, the point from which your measurements will be made. The orientation of your telescope will determine whether that angle is clockwise or counterclockwise. Check this by moving the telescope's declination axis north. This will be 0 degrees, with east being 90 degrees.

We continue setting up by determining the number of arc seconds in one rotation of the micrometer tangent screw. This value is found by dividing the difference in declination of two known stars by the number of screw revolutions. Ideally, the procedure is quite simple; you are working backwards by taking a star of known separation and determining the number of arc seconds in each screw turn. Once that is achieved, you then use these known values to learn about stars whose separations and position angles are unknown to you.

To measure the position angle of a double, set one wire so that it crosses both the primary and the secondary stars. Keep your eyes parallel to the line joining the stars, so that your reading of the position is accurate. Then you simply read the angle. Repeat the measurement several times.

To measure the separation of one component from the other, place the fixed wire over the primary star, and use the micrometer screw to move the second wire to the position of the secondary star. Using the screw's scale, which you have already calibrated, record their separation in seconds of arc.

Obviously, the filar micrometer is a far more accurate method of measuirng change in binary systems. However, the simpler star atlas method should show results as rapidly orbiting stars show more definite change.

# 16   Variable stars

In the thrilling and captivating fluctuations of the variable stars lies a field where an amateur can aid significantly the advance of knowledge. Variable stars are simply objects whose light output is not constant, and your goal as an observer of these stars is to determine their changing brightnesses as accurately and as objectively as you can.

For every observing area we have explored so far, the one constant is change. The planets offer their shifting atmospheres, surface features, and satellites; the comets display their movements and the changes of their comae and tails; and the Sun shows its daily march of spots. Now we enter a realm where change, though not obvious over a night, is what keeps up our interest.

When we look at a variable star we see only one point of light amidst a field of lights, and in fact the first problem is only to find which star is the variable. With variable stars, the thrill is not instantaneous; it lies in their changing from night to night, or week to week. It grows on us.

## 16.1   The AAVSO

What, you might ask, is the value of your making these observations with a back yard telescope? If you were the only person making these observations, you may actually have come up with some startling new facts about the behavior of variable stars, but not much else. The real value of visual observations lies in the fact that your work is supported by that of hundreds of other people, who like you have small back yard telescopes and are watching these stars from observing sites all over the world. Together, the many observations made worldwide form a very useful dataset.

Most of these hundreds of observers report to the American Association of Variable Star Observers (AAVSO), an organization that has been fostering the observation of these stars and collecting data since 1911. Its address is 25 Birch Street, Cambridge, Massachusetts, 02138, USA. The AAVSO is one of the most highly respected amateur organizations in the world. Founded for a single purpose, the encouragement and archiving of variable star observing, this organization maintains a staff of people who record and archive the many observations, placing them in a form that will be useful to those who wish to learn about the behavior of variable stars.

The AAVSO consists of extraordinary people who have found a special motivation for their astronomical commitment. These are people who began, like you and me, with small telescopes, looking at the Moon, double stars, and deep sky objects. Some of these people, but probably a minority, have found such satisfaction in variables that they observe nothing else. Most of the variable star observers today are also the deep sky people who show those stunning photographs at club meetings, the avid planetary people, and the volunteers for public star parties. The AAVSO, with its specific and challenging programs, provides direction to the observing activities of these people.

Janet Mattei directs the AAVSO and maintains its file of some five million observations to standards acceptable to the professional community. A professional astronomer, Janet Mattei is familiar with all branches of variable star astronomy and with which stars astronomers are most concerned from one year to another. By encouraging members to concentrate on these stars, their work becomes even more important.

Not all the observations you may make will see the light of professional study, and you may never know if they are; it is also impossible to know, when you are making an observation, if someday it will come to be useful. But that's not the point. The 20 or 30 observations you may make each month, combined with those of other observers, will help form a picture of a variable's fluctuating brightness; we call this picture a light curve. Your observations can improve the light curves of the stars in your program. Astronomers requesting AAVSO data will naturally use those stars for which the most observations exist, and in that sense any observation you make is helpful.

With most of their observers striving for the most careful and accurate estimates to one tenth of a magnitude, a light curve can be produced for a star that is accurate to an uncertainty of about 0.25 magnitude. An astronomer looking at a set of data will get far more out of it if its accuracy is not exaggerated. Actually, the accuracy of visual observations is not as good as astronomers would like, but it is certainly good enough to give them an idea of what stars are doing interesting things and need to be followed up with professional instruments.

Professional astronomers, with their photometers and charge-coupled devices (CCDs), can produce magnitudes accurate to about one hundredth of a magnitude. But these instruments are by no means available in such massive numbers and worldwide locations as the visual network. So it is up to us to watch lots of stars, report on them regularly to the AAVSO, and to be alert for unusual activity. What if, for example, SS Cygni suddenly rises from 12th to 8th magnitude in a single night? At maximum, the star produces X-rays which can be observed by satellites like the Einstein Observatory. To plan their observing program, the Einstein observing team relied on amateur observations to tell them when cataclysmic variables like SS Cygni were entering outburst. The highly successful International Ultra-violet Explorer (IUE) and Hipparcos satellites also have made use of visual observations in their mission plans.

## 16.2 Eclipsing binaries

In the southwest corner of the parallelogram of Lyra is a variable star called Beta Lyrae. Its brightness can be compared with two nearby stars in the parallelogram, a bright one called Gamma Lyrae to the east, and a fainter one called Zeta Lyrae to the north. The magnitude of Gamma is 3.3, and Zeta is 4.3. These magnitudes do not change over time.

How bright is the apex star relative to the two others? Say it is about halfway between. That would mean its magnitude is 3.9. And thus your variable star estimate is made!

Still not the most exciting thing? Maybe not; all you have done is to determine the brightness of one star relative to two others. But in two or three nights if you try the exercise again, you will find that Beta Lyrae has changed. Instead of 3.9, it may have brightened to 3.7 or faded to 4.1. Change has occurred before your eyes over two nights! Now is the exercise getting exciting? The thing about variables is that as we estimate these stars we are witnessing the change going on within them.

Stars change in brightness for all sorts of reasons. Beta Lyrae is an eclipsing binary; in fact, not a real variable at all, even though its light output does change. It varies because of a coincidental lineup of two orbiting stars and the Earth. When this happens, one member of a distant binary system will periodically eclipse the other one. When that happens the light from the whole system drops; we can see and measure that.

## 16.3 Cepheids

The Cepheid variables are similar in that their periods can be determined precisely, although the mechanism is very different: Cepheids vary because of intrinsic physical changes going on in the stars. The star we chose for our example back in Chapter 2, 'Without a telescope', was Delta Cephei, type-star of the whole class. In 1910 Henrietta Leavitt studied about 25 stars in the Small Magellanic Cloud, one of the two nearest galaxies to the Milky Way, and learned that the longer a star's period of variation, the brighter its average brightness. Realizing that all the stars Leavitt studied are roughly the same distance from us, Harlow Shapley later developed the important period–luminosity relation. If we observe a Cepheid that varies from 12th to 13th magnitude, we know it is farther away than one that varies from 9th to 10th magnitude if its period of variation is the same. This discovery gave us our first accurate method to determine the distances to globular clusters and nearby galaxies.

This historical impact is not the reason we as amateurs observe Cepheids now; we watch them to see their cycles of change. Variable star observing, like fine wine, is an acquired taste; you won't be hooked the first night.

## 16.4  Long period stars

Some red stars with about the Sun's mass, but much larger, change over leisurely cycles of several months. They also vary over huge magnitude ranges; Chi Cygni, for instance, drops from a maximum magnitude of 5.2 to a minimum of 13.4, a variation of eight magnitudes! Also known as Mira stars, after the first known example, these are the most common type of variable. Variations of several magnitudes can occur in leisurely intervals of about 200 to 400 days, and you should check on their progress every two weeks.

Mira stars form the backbone of the visual contribution to the study of variable stars. Their leisurely changes, several-magnitude differences between maximum and minimum, with periods lasting many months, make them ideal candidates for back yard telescopes. With that type of variation, magnitude estimates to one hundredth magnitude are not really necessary, and their sheer numbers make it impossible for professional astronomers with limited telescope time to cover them with any depth. Accordingly, until about 1970, when eruptive stars became popular targets, long period stars received by far the most attention from amateurs. They are still important stars to observe, and any professional variable star astronomer who studies Miras will make use of visual data.

Two Miras are my personal favorites. One is R Leonis, whose 5.9 maximum makes it an easy binocular object when it is bright. It forms a tight triangle with two ninth magnitude stars, and its red color is quite striking. It drops to a minimum that averages magnitude 10.2. The other is V Bootis, which is easy to find by moving about 0.75 degree northwest of the third magnitude star Gamma Bootis. V Bootis is always bright enough to see with a small telescope, whether it is near its 7.5 magnitude maximum or towards its minimum at magnitude 10.7. Even though V Bootis is classified as a Mira star, it does undergo some interruptions in its cycle that can last for some time, so it is always worth watching. Figure 16.1 is a chart for V Bootis. Designed for finding the star and for binocular viewing when the star is near its maximum, this chart shows north up. Figure 16.2, with north down, is designed for telescope use when the variable is closer to minimum brightness.

Figure 16.1. R Canum Venaticorum and V Bootis. Chart designed for unaided eye or binoculars. Reprinted by special permission of the AAVSO, through its Director, Janet Mattei.

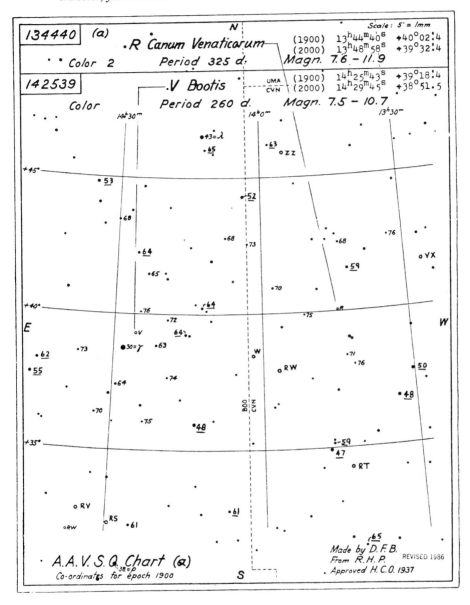

Figure 16.2. V Bootis. Chart designed for the inverted field of a Newtonian reflector. Reprinted by special permission of the AAVSO, through its Director, Janet Mattei.

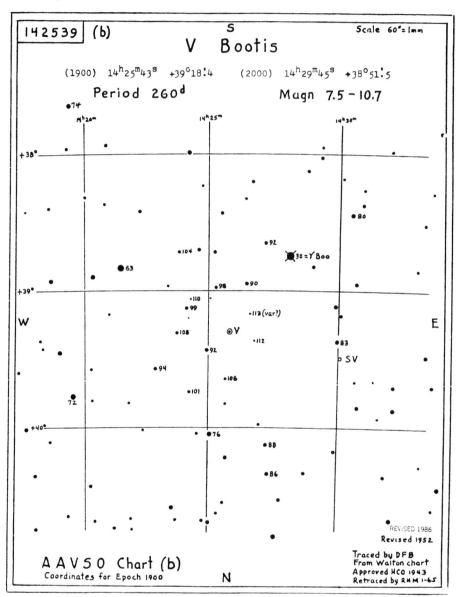

## 16.5   Semiregular stars

One of the brightest stars in the sky, Betelgeuse, is also an interesting variable star. Betelgeuse is a supernova candidate, but don't worry yet: it is only 100 light years away, but it may be millions of years before Betelgeuse does anything frightening. But right now, tonight, Betelgeuse is doing something interesting – it is being a semiregular variable star.

Although its period is almost seven years, you can sometimes observe changes in this star within a week or two. Estimate twice a month, and use stars that do not differ too much in altitude from the variable since atmospheric extinction will introduce false differences in magnitude. Also, since Betelgeuse is a red star, use quick glances rather than prolonged stares to avoid the red buildup on your eye.

Another bright red giant that varies irregularly and is worth a look is Eta Geminorum, not far from the open cluster M35 in Gemini. This star varies from 3.2 to 3.9 in a period of about eight months. It is easily found and estimated with unaided eye or binoculars. Other variables are not so regular. Some bright red stars, often visible using binoculars, vary by small amounts over shorter but less certain periods.

## 16.6   Cataclysmic variables

Where the Mira stars often display uncertainty in the details of their periods, the cataclysmic, or eruptive, variables are often completely irregular.

Eruptive stars vary irregularly and explosively. Perhaps the best known is SS Cygni, a 'dwarf nova' whose outbursts from 12th to almost 8th magnitude have been assiduously watched since 1896. These stars are binary systems in which hydrogen transfers from a large star to its smaller neighbor. When too much hydrogen surrounds the smaller star, it is simply blown away in an explosion. Their penchant for unexpected behavior has drawn many observers to their fold. Recently discovered X-ray emissions, using orbiting satellites, have generated intense interest among some professional astronomers. Thus these stars are really a top priority observing project for amateurs.

SS Cygni is the most popular example of what is known as a dwarf nova, in which the transfer of matter from one star to another causes a small explosion every 50 days or so. The uncertainty in the period is only one of two factors that makes such a system exciting to watch. In fact I have observed eruptions separated by only two weeks and another time I watched SS Cygni slumber peacefully as three months dragged by.

The other question concerns the nature of the eruption. Some outbursts are sudden, and you can watch in wonder as a tiny 12th magnitude point of light brightens in a few hours by almost four magnitudes. Others take more time. Once I watched SS Cygni for a whole week while it struggled along to maximum, hovering every now and then at 'pit stops' of 10.6 and ninth magnitude along the way. You can make useful observations of SS Cygni even with a 7.5 cm telescope, even though you really need a 15 cm or more to catch it all the way to its 12th magnitude minimum.

Stars like SS Cygni do not necessarily have to be followed at minimum. If you see nothing where SS Cygni should be, and the faintest star you see is 10.8, simply record the estimate at (10.8, the '(' meaning that SS Cygni is fainter than 10.8, which was the faintest star that you could see in the field that night. Then, if the next night's observing reveals a new tenth magnitude speck of light, you are as aware as anyone that SS Cygni is about to go into outburst. When professionals are actively working on the dwarf novae, they need the constant attention that our amateur network can provide.

A rare form of eruptive star is R Coronae Borealis, which undergoes completely irregular and sudden *drops* in brightness from its usual sixth magnitude maximum. Although the star is eruptive, the material it erupts is like soot, which masks the star's light, causing it to fade. Over the course of a few nights R Coronae Borealis may drop as much as three or four magnitudes. Sometimes the star drops to almost the 15th magnitude.

## 16.7  T Tauri

In the starlit heart of the Hyades cluster lurks T Tauri, the best known of a special class of young stars that vary irregularly, offering both long term variations and short term changes, neither of which can be predicted. T Tauri is a star which should be estimated at least once per night, and if your session lasts a long time, try every hour. You may notice subtle changes up to a quarter of a magnitude, if you are lucky.

## 16.8  Naming of variables

Variable stars are named according to a formula that began as the brainchild of Father Argelander, one of the early great variable star observers. Not wanting to confuse his scheme with the lower-case Bayer letters a through q that had been in use as star names since 1603, Argelander began by calling the first star to be discovered as variable in a constellation as R, the second S, until Z. (T Cephei is an example.) As variable discoveries became more frequent, designations from AA to AZ were adopted (AR Cephei), followed by BB to BZ (BM Orionis), CC to CZ (CH Cygni), and so on. The letter J was omitted. Finally, in one constellation the list ran to QZ. After that, the next variable was listed as V335. V1500 Cygni is the name for the bright nova that appeared in Cygnus in 1975.

An independent system, known as the Harvard designation, was adopted almost a century ago, using the 1900 co-ordinates of a star. Thus, 213843 SS Cygni refers to SS Cygni's 1900 position of 21 hours, 38 minutes, plus 43 degrees, and 021403 signifies Omicron Ceti's position (the star is also called Mira) of 2 hours, 14 minutes, and *minus* 3 degrees. The last two digits, 03, are underlined to indicate south declination.

Still a bit confusing? Adapting my 'Naming of Comets' that was adapted from T. S. Eliot's 'The Naming of Cats', and which appears at the end of Chapter 14, Steve Edberg came up with 'The Naming of Variables' (with apologies to T. S. Eliot):

> The naming of variables is a difficult matter,
> It isn't just one of your holiday games;
> You may think at first I'm mad as a hatter
> When I tell you, a variable has THREE DIFFERENT NAMES.
> First of all, there's the name that the family use daily,
> Such as Eclipse, Symbiotic, Cepheid, or Flare,
> Such as U Gem, R Cor Bor, Z Cam, T Tauri –
> All of them sensible, everyday names.
> There are fancier names if you think they sound sweeter,
> Some for the gentlemen, some for the dames:
> Such as Omicron Ceti, V Canum Venaticorum,
> RU Lupi, UZ Serpentis, T Piscium,
> T Coronae Australis, Beta Persei,
> But all of them sensible, everyday names.
> But I tell you, a variable needs a name that's particular,
> A name that's peculiar, and more dignified,
> Else how can he keep up his light variation,
> Or spread out debris clouds, or cherish his pride?
> Of names of this kind, I can give you a score.
> UV Ceti, SS Cygni, SU Tauri and such,
> Or T Pyxidis, R Bootis, S Persei – there's more –
> Names that never belong to more than one star.
> But above and beyond there's still one name left over,
> And that is the name that at first you can't guess;
> The name that no human research can discover –
> Until long after the variable's brightened and faded,
> Like Nova Cygni, '75 –
> But the VARIABLE HIMSELF KNOWS, and won't now confess.
> When you notice a star in profound meditation,
> The reason, I tell you, is always the same:
> His mind is engaged in rapt contemplation
> Of the thought, of the thought, of the thought of his name:
> His rotational, orbital,
> Nuclear reactional
> Deep and inscrutable singular Name.

## 16.9  How to observe a variable star

1. Find the variable. Sounds easy, but often this is the most difficult part. If the field of view is a new one for you, you could spend an hour or more finding the star. If the star is at minimum it could be too faint for you to see in any case!

It helps to have an idea before hand if a star will be bright enough to see. If I suspect that a star will be too faint to find in my telescope, I will leave it alone, unless the star is part of a field I know well.

2. Choose two comparison stars, one which is brighter and the other fainter than the variable. You find the field of a variable, including the comparison stars and their magnitudes, from a series of star charts that you get from the AAVSO. We have already done this exercise with Delta Cephei and Beta Lyrae.

3. As you proceed through a night's observing, you will observe a number of different stars. It is best to record these directly into your observing log so that they will not get lost. Do not record them in any way that will allow you to see what your last observation was, as you do not want to be influenced by it.

4. When you get inside after your session is over, copy each observation into a list you have prepared for each star. You may use a set of index cards with one card for each star, or a computerized database from which you can examine all the observations for a particular star. More sophisticated programs might automatically insert a night's observations into the proper files. Whatever method you use, the important thing is to keep your list up to date as the nights go by.

5. At the end of the month, fill out the report to send to AAVSO. They ask that you order your observations by star, and then chronologically for each star, and if you have been doing this all along anyway your work is easy. The AAVSO likes to have their reports as soon as possible after the first of each month, and this is a good idea for its forces you as observer to prepare a report while your memories of the past month's work are still fresh.

**Some hints** A. Variable stars tend to be red, and some stars are *very* red. Unfortunately, red light has a tendency to build up on your retina, making such an object appear brighter than it really is. In fact, the red light builds up on your eye like light on photographic film, so that the star may appear to brighten as you watch it. This is known as the 'Purkinje' effect, and the best way to avoid it is to look at the star with 'quick glances' rather than with long relaxed stares.

B. Just because some stars can be seen with binoculars or even the unaided eye, don't assume that these brighter ones are easier to estimate. They can be very difficult, and it is actually better to begin a variable star program with the fainter, telescopic, stars.

C. For professionals interested in Mira star data, our variable star observing the visual way is almost unacceptably subjective. As photometers in the visual spectrum, our eyes are far from ideal instruments since they are connected to our subjective brains. Too often we may make the mistake of basing an estimate on our recall of what the star looked like one or two weeks ago.

To help reduce this level of uncertainty, I suggest that you record your observations twice. Out in your observatory, or at your telescope, keep a notepad with a blank sheet of paper on which you record your variable estimates. Do not prejudge your work by looking at your last value. Then, inside, after your session, copy your estimate underneath all the other magnitude estimates for that star during the month.

## 16.10 Suggested frequency of observation

**Long period or Mira** These stars vary slowly, so observe them every two weeks unless you think the star is near minimum or maximum. At these two more-critical points of a star's variation, estimate once every week or even every few days.

**Semiregular stars** These stars usually show less activity than the Miras, so estimate once a month; certainly no oftener than every two weeks.

**Eclipsing binaries** For stars with sharp minima, like Algol, make several estimates per hour while minimum is in progress. With these stars, the key is to know beforehand (from the AAVSO's Eclipsing Binary Division) when stars are going into eclipse.

**Cepheids** Depending on length of period once a night to once per week.

**Cataclysmic variables** These stars are not highly predictable, so an estimate every night is useful. If you see a 'dwarf nova' going into outburst, follow its progress by estimating every hour. If you have watched a star each clear night, and after many nights of hovering around minimum the star is now a half magnitude or more above minimum, then you should assume the star is erupting and increase your frequency of estimates. Estimate R Coronae Borealis stars once a night also. You never know when stars like this will begin their plunges to minimum!

## 16.11 Northern summer program

Using Figure 16.3, the chart for 155947 X Herculis, try to form a picture of the behavior of three interesting variables over the months of summer: X Herculis, g Herculis and RR Coronae Borealis. Estimate each variable twice a month, and later, perhaps with the help of the estimates of other people, you can determine a light curve for these stars.

Figure 16.3. X Herculis and friends. Chart designed for unaided eye or binoculars. W Herculis and W Coronae Borealis are Mira-type stars; the other three are semiregular. Reprinted by special permission of the AAVSO through its Director, Janet Mattei.

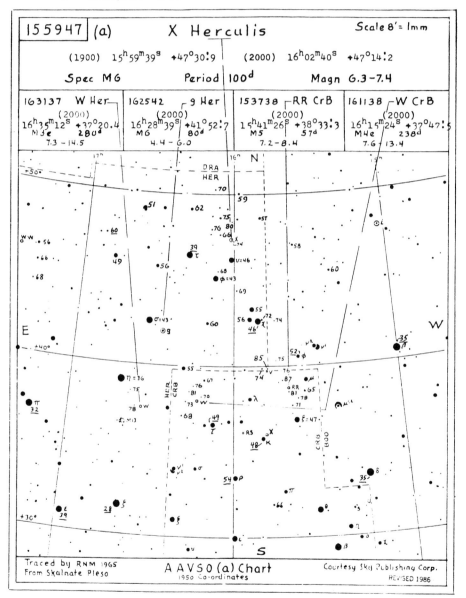

# 16.12 Northern winter program

Try the same thing with Figure 16.4, the chart for 050001 W Orionis. Also appearing on this chart is RX Leporis, a star with some peculiar variations. Estimate each twice per month.

The beauty of these programs is that they are challenging, but not too difficult. Try them!

Figure 16.4. W Orionis and semiregular neighbors. Chart designed for unaided eye or binoculars. Reprinted by special permission of the AAVSO, through its Director, Janet Mattei.

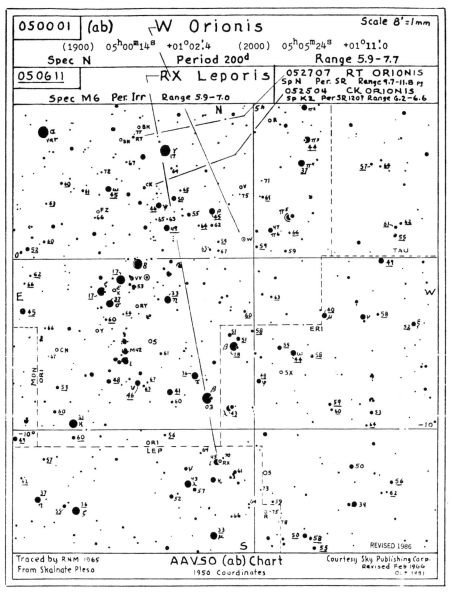

Figure 16.5. Mira and R Ceti. Chart designed for unaided eye or binoculars. Reprinted by special permission of the AAVSO, through its Director, Janet Mattei.

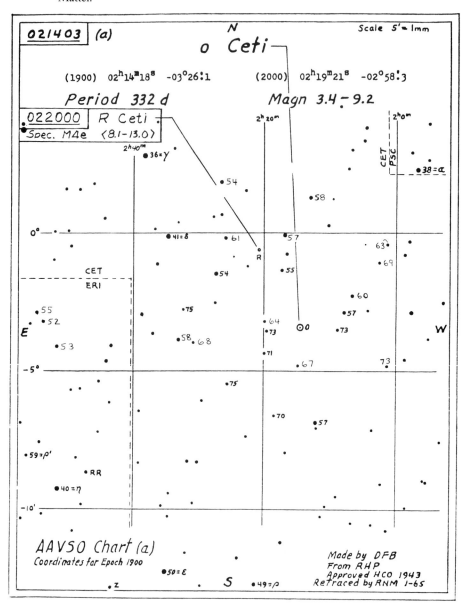

Figure 16.6. Mira. Chart designed for the inverted field of a Newtonian reflector. Reprinted by special permission of the AAVSO, through its Director, Janet Mattei.

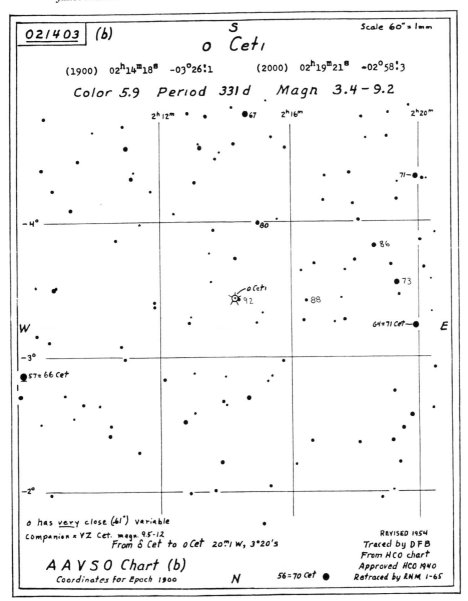

## 16.13 A selection of variable stars

The following list of selected stars comes from the AAVSO catalog. The stars are listed by their Harvard designations, their names and their magnitude ranges. I suggest you specify a complete set for each star when ordering these charts from the AAVSO. Underlined designations indicate that the star is south of the celestial equator.

021403 Omicron Ceti (Mira)
2.0–10.1, 332 days.
The type-star of the long period variables. See Figures 16.5 and 16.6.

021558 S Persei
7.9–11.1.
A good semiregular variable.

043274 X Cam
8.1–12.6, 143 days.
At 143 days, one of the shortest of the long period variables. Also circumpolar.

045514 R Leporis
6.8–9.6, 432 days.
One of the reddest stars in the sky; a delight to watch.

054907 Alpha Orionis
0.4–1.3, 2070 days but not regular.
Did you know that this magnificent star also was slightly variable?

064030 X Gem
8.2–13.2, 263 days.
Long period variable.

081473 Z Cam
10.2–14.5.
A dwarf nova that sometimes gets 'stuck' on its way down from maximum; has been observed to stay almost constant for years!

094211 R Leonis
5.8–10.2, 313 days.
A classic long period variable.

122402 3C273 Virginis
12.0–13.2.
Quasar, slightly variable.

123160 T Ursae Majoris
7.7–12.9, 257 days.

123459 RS UMa
9.0–14.3, 260 days.

123961 S Ursae Majoris
7.8–11.7, 226 days.
Three stars in almost the same field.

142539 V Bootis
7.0–11.3, 258 days.
Some less than regular fluctuations. See Figures 16.1 and 16.2.

154428 R CrB
5.8–14.8.
A special star that stays at maximum for indefinite periods, then plunges down for a while to its minimum.

155526 T Coronae
Borealis                    2.0–10.3.
A recurring nova. Last erupted in 1946. See
Figures 16.7 and 16.8.

Figure 16.7. R Coronae Borealis, RU Herculis, and T Coronae Borealis. Chart
designed for unaided eye or binoculars. Reprinted by special permission of the
AAVSO, through its Director, Janet Mattei.

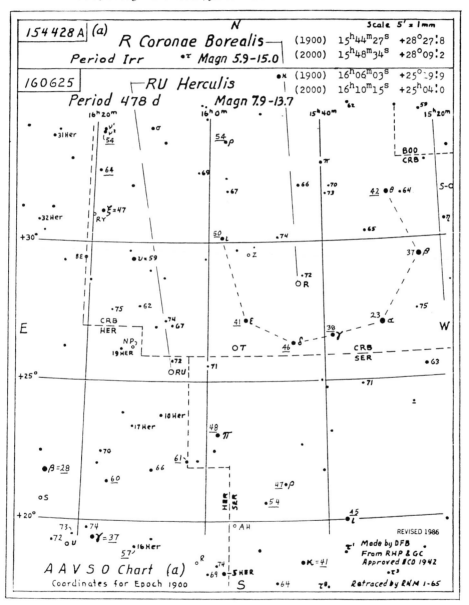

174406 RS Ophiuchi        5.3–12.3.
                          Also a recurring nova. Estimate once a night.
194632 Chi Cygni          5.2–13.4, 407 days.
                          A famous Mira star, but can be quite difficult
                          when faint.

Figure 16.8. T Coronae Borealis. Chart designed for the inverted field of a
Newtonian reflector. Reprinted by special permission of the AAVSO, through its
Director, Janet Mattei.

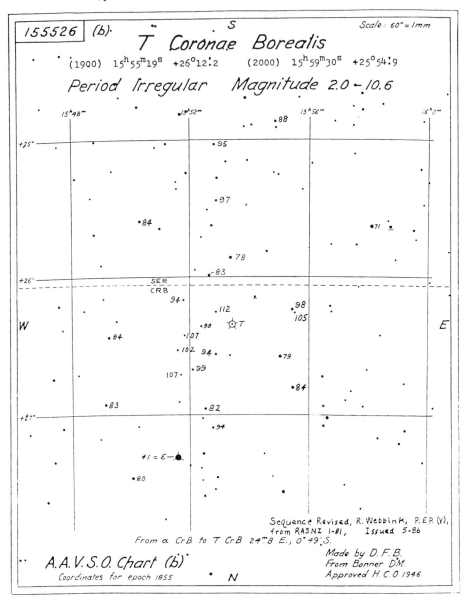

| 201520 V Sge | 9.5–13.9. |
| | Nova-like and unpredictable. |
| 214058 Mu Cep | 3.6–5.1. |
| | Irregular and very red. Herschel called this the 'Garnet Star'. |
| 220912 RU Pegasi | 9.0–13.1. |
| | Another SS Cygni-type dwarf nova. Average period 66 days. |

## 16.14 Searching for novae and supernovae

My first view of a nova left me in a state of utter disbelief. Looking at Cygnus, I saw a dazzling new star just north of Deneb; to look up at the sky and find it right along the dark arm of Cygnus was a thrill. Later I learned that the star had been bright the night before, but clouds over Montreal prevented its appearance here. All through that first night the nova seemed to stay at about magnitude 1.7. The next night I drove out of town to get a better view of the stranger, but already it had dropped by half a magnitude.

Searching for novae, or stars that suffer explosions in their outer layers, is a field that belongs to the amateur. With unaided eye, or binoculars, you can search parts of the sky for novae or supernovae. The only prerequisite is a thorough knowledge of the constellations, and of all the stars down to whatever magnitude limit you are searching.

The AAVSO has a Nova Search Division in which the nova-prone Milky Way and its vicinity is divided into over 100 small regions. Like the comet search areas, you search your region to a certain limit, usually from seventh to ninth magnitude. Anything you do not remember must be checked out, but almost every suspect will turn out to be a star that belongs there. Remember that even negative searches have value. What if, for instance, a nova is discovered in your area on March 15, but your March 14 report shows nothing unusual above sixth magnitude? Then your report helps to pinpoint the upper limit of brightness for the nova on the night before discovery.

Peter Collins, discoverer of three novae between 1978 and 1987, divides the regions he searches with binoculars into patterns that he can identify. Once he knows these patterns he recognizes instantly any star that appears out of place.

After you get to know the sky well, nova finds can be almost automatic in the sense that even a passing satellite should alert you to 'something wrong' in the sky. And if you do find a nova you must notify the Central Bureau for Astronomical Telegrams, just as for comets. They need the nova's position and brightness, as well as the time of the observation.

The major stellar explosions called supernovae can outshine the combined light of the whole galaxy in which they reside. Since no supernova has been seen in our galaxy since 1604, searchers concentrate on every available galaxy in their search for new 12th to 14th magnitude stars. Identifying a new supernova is quite difficult, since foreground stars (in our own galaxy) and starlike 'H2 regions' in many galaxies are often confused with supernovae. A handful of amateurs were searching for supernovae in distant galaxies before Reverend Robert Evans of Hazelbrook, New South Wales, began finding one after another. By becoming familiar with and searching every galaxy he could see with a 30 cm reflector, and later a 40 cm reflector, he began discovering them so frequently it seemed easy.

Actually, it wasn't. By the end of 1987, Reverend Evans had discovered 16 supernovae, a major achievement that has allowed astronomers to

Figure 16.9. The Large Magellanic Cloud, with Supernova 1987A near maximum brightness. The supernova is the bright star to the right of the galaxy. The object immediately to the right of the supernova is the Tarantula Nebula, a large diffuse nebula. Steve Edberg took this five minute exposure on May 29, 1987, with a 260 mm lens at f/5.2, on 3M1000 film.

increase their understanding of these important stars. During that year Evans made 15 554 observations of galaxies with two discoveries taking place. If a galaxy-rich region of the sky, like Leo and Virgo, is prominent, on some good nights Evans might average about three to five hours of searching, in which time he would check some 200 galaxies. His record for a single night of searching at the time was 570 galaxies. In setting such a phenomenal record for a visual search for supernovae, he has created a new competition, just as Messier developed the art of hunting for comets two centuries ago.

At the end of February, 1987, three observers independently discovered a supernova in the Large Magellanic Cloud, one of the nearest galaxies to our own. Ian Shelton found the interloper's image on a photographic plate he had taken, and Oscar Duhalde, from almost the same site, observed the star

Figure 16.10. Detail of previous exposure showing Supernova 1987A, with Tarantula Nebula to its lower right. Photograph by Stephen J. Edberg.

with his unaided eye. Finally Albert Jones, an amateur astronomer from New Zealand, discovered the exploding star during an observing session.

Known as Supernova 1987A (the first supernova to be discovered in 1987), this was the brightest supernova to be seen from Earth since 1604. A few weeks after the discovery, Steve Edberg, Charlie Oostdyk, and I travelled to Acapulco, Mexico, in hopes of observing this singular event. On the first of two possible nights we missed seeing the object as it set behind a cloud. The second evening we hired a taxi to go deep into the interior of southern Mexico to find a better site. Although the sky was clear, our site was not; we watched helplessly as the supernova remained hidden behind a nearby hill. We raced to a different site, and even crawled under some barbed wire and raced up a hillside, telescope in hand. Although we were too late for the supernova, we did get a good view of Eta Carinae, a most unusual variable star, from that 'illegal' site.

Departure day was clear and sunny, so we decided to extend our trip. That evening we set up on a cliff overlooking the ocean, and by 'star-hopping' from one star to another, comparing them with our map, we finally saw, just above the horizon, a bright red star. Shining at us from a distance of some 160 000 light years, Supernova 1987A was an extraordinary sight. We had observed a star from another galaxy, estimated its magnitude at 4.4, and returned home the next day. The cost of this single variable star observation, including our $75 taxi fare on the second night, was about $600 per person!

## 16.15 Neutron star song

At the center of what is left after a type II supernova is a spinning neutron star. Shall we celebrate such a star in song? Try *Neutron Star*, sung to the tune of Loretta Lynn's *Jealous Heart*, an old country and western song:

Neutron star, oh neutron star, you're massive
And your tidal forces are intense.
You have crushed your atom shells to pieces;
Neutron star, your gravity's immense!

You were once a star like all the others,
Shining brightly in the evening sky
Till your thermonuclear reactions
Consumed all your hydrogen supply.

Neutron star, oh neutron star, you're spinning
Round and round at such a fever pitch.
You conserve your angular momentum
And speed up at every little glitch.

You were once a star like all the others
Somewhere on the Hertzsprung–Russell graph;
Now you're in the lower left-hand corner,
Stellar mass reduced by almost half.

Neutron star, oh neutron star, you're pulsing
Twisting your magnetic lines of force,
And electrons spewing from your axis
Form a synchrotron emission source.

You were once a star like all the others
Till your hydrostatic balance failed,
And you lost your radiation pressure
And your outer chromosphere exhaled.

Peter Jedicke, 1981

# 17  The deep sky

Past the planets, and beyond the double stars and the variables, is a special realm. Here are the star clusters, gas clouds, and the galaxies that are traditionally known as the 'deep sky objects'. With few exceptions, these objects cannot be seen without binoculars or a telescope, and to do most of these objects justice, you should see them under a dark sky far away from the pollution of city lights.

The concept of deep sky objects is a complex one that has evolved from at least the seventeenth century. Edmond Halley was one of the first observers who, with a telescope, recorded a number of cloudy objects that eventually became known as 'nebulae'. We have already read of Charles Messier, who in 1758 observed Halley's Comet and who subsequently kept a list of the fuzzy objects that confused his search for comets.

The 110 objects in Messier's catalog are interesting enough to deserve a separate chapter, so we will discuss them later. But, just for now, let's say that the Messiers are a mere sampling, though an admittedly thorough one, that you should examine as you begin to explore this mysterious and distant kingdom.

## 17.1  The *New General Catalogue*

The *New General Catalogue* was prepared by an astronomer who had first gained a reputation as a biographer of Tycho. John Louis Emil Dreyer published his list in 1888, partly to attempt to combine a diverse set of catalogs then in existence. These included a list of about 2500 nonstellar objects that had been completed by William Herschel in 1802, and his son

John Herschel added more from his southern hemisphere studies. By 1863, the Herschel catalogs described 5096 objects. In 1877, Dreyer published a list of 1136 objects that he, d'Arrest, and others had observed.

The idea for a general catalog emerged when Dreyer attempted to publish a supplement to his own work. Herschel's *General Catalogue of Nebulae* was no longer easily available, and the Council of the Royal Astronomical Society suggested that Dreyer produce a catalog incorporating the information in all these lists. Finally, in 1888, Dreyer published Memoir 49 of the Royal Astronomical Society, 'A new general catalogue of nebulae and clusters of stars, being the catalogue of Sir John Herschel, revised, corrected, and enlarged'.

There are 7840 objects in the *New General Catalogue*, which might seem like a lot, but other catalogs add many more. By 1908 Dreyer himself had added over 5000 in his *Index Catalogue*. This list merges types of deep sky objects that are intrinsically very different, from illuminated masses of dust and gas to the huge masses of stars and other material that are the distant galaxies. The only thing that keeps all these things in one list is their appearance – they tend to look like fuzzy patches of light. The index catalogue objects, it should be noted, were discovered mostly on the photographic plate and are very faint. IC434 in Orion, the bright nebulosity that surrounds the dark Horsehead Nebula, is an example.

## 17.2  Open clusters

These are groupings of stars that move together through space, and were formed together out of huge clouds of hydrogen. In Sagittarius we can actually witness this stage of cluster development; the open cluster NGC6530 is associated with the Lagoon Nebula M8. Of older clusters, the Pleiades is by far the best known example. It is so prominent because its 500 stars are relatively close. Just south of the Pleiades is an even larger cluster called the Hyades. One of the closest clusters to us, the Hyades spreads out over several degrees of sky, and is punctuated by bright-red Aldebaran, an interloper that is not a member of the cluster.

In the coarsest clusters the stars which are far apart are often barely distinguishable from the stars around them. One is a group of stars at about half the distance of the Hyades; it is so close, that you probably wouldn't recognize it as one. It consists of some of the Big Dipper's bright stars, including Epsilon, Beta, Zeta (Mizar and Alcor), Gamma and Delta. Alpha Coronae Borealis may also be a member. The center of this cluster is only some 75 light years away. This cluster seems to be surrounded by a group of other stars which, though sharing the speed and direction, may or may not belong to the cluster. This group, sometimes called the Ursa Major Group or Ursa Major Stream, consists of such prominent stars as far around the sky as Alpha Ophiuchi, Delta Leonis, Beta Aurigae, and even Sirius. The Ursa Major Stream virtually surrounds us. In fact, although the Sun is not a member of this group, it is currently passing through it! It is unknown whether there is any real connection between the Ursa Major Cluster and its surrounding stream.

There is a difference between chance groupings of stars which we see scattered all over the sky, and genuine open clusters. Membership in a real cluster depends on a star's sharing two factors with already established members. These factors are proper motion and radial velocity.

Early in 1929, Clyde Tombaugh began a photographic patrol program at Lowell Observatory designed to find a new planet. He succeeded less than a year later, but fortunately he and his colleagues at Lowell Observatory decided to continue the search. For 13 more years the program continued, resulting in a beautiful archive record of most of the sky. A quarter century after the first program began, the observatory decided to photograph the sky again, using the same photographic camera, but this time to measure the displacements of some of the stars. This 'proper motion survey' was highly successful in giving us a picture of the motions of stars in our corner of the galaxy.

What is proper motion? Since 1718, when Edmond Halley noted that Arcturus and Sirius had moved slightly relative to positions Ptolemy had noted some 12 centuries earlier, we have been aware that the 'fixed' stars actually move relative to one another. These motions are known as proper motions. Most stars are so distant and appear to move so slowly that 50 to 100 years may be required to show this motion.

Figure 17.1. M103, an open star cluster. Photo by Tim Hunter of Tucson, Arizona. 60 cm. f/5 reflector.

In addition to these slight displacements, we need to know the line-of-sight velocities of stars, how fast and whether they are moving away from us or towards us. These are known as radial velocities. If a star's proper motion and radial velocity is identical with that of other nearby stars, that would indicate that the stars all belong to a cluster. We do need other data to confirm membership. The star obviously has to share the same distance as the other members, and since open clusters consist of stars that share a common genesis, the age should be the same. A star moving differently from the established members could be a foreground or background star or, if it has the same distance from us, merely be an interloper passing through the cluster.

Open clusters were classified by Harlow Shapley according to their concentration of stars, where

> c = very loose and irregular,
> d = loose and poor,
> e = intermediate rich,
> f = fairly rich,
> g = considerably rich and concentrated.

Besides the tightly bound clusters like the Pleiades and the Beehive, there are also looser assemblages of stars called associations. Embedded in the Orion nebula are hundreds of faint blue stars that form what we call an association. First noted by the Russian astronomer Viktor Ambartsumian, these associations are not held together very strongly by gravity and therefore do not last more than half a billion years or so, which cosmologically is quite a short time.

## 17.3  Globular clusters

Where open clusters consist of hundreds of stars, globular clusters involve hundreds of thousands, and in the cases of very large clusters, a million or more stars. Fuzzy blobs of light through a small telescope, these lovely objects will continue to enthrall you no matter how much you increase the power of your telescope; as you look through larger and larger telescopes the view will keep on getting better.

My first view of M13, the globular cluster in Hercules, was through a tiny 3.5 inch reflector. A generation later an eight-year-old girl would say to me that it 'looked simply like a tiny fuzzy animal'. A few months later I had another first look, this time through a 40 cm reflector with a new 9 mm Nagler eyepiece that offered a combination of wide field and high power. I was busy hunting for comets, so I did not want to spend more than a few minutes looking at M13. When I first saw the cluster taking up most of the field, resolved right through to the center as a sea of stars interspersed with dark lanes, I was so taken aback that I forgot all about comets. The fuzzy animal had grown up fast!

Globular clusters are important for a number of reasons. They are so large and distant that determining their distances can begin to give us an idea of the size of the universe. Second, globulars are among the oldest members of our galaxy, so studying them can give us an understanding of galactic evolution.

Some 131 globulars in and around our galaxy have been identified, ranging in angular size from Omega Centauri, 47 Tucanae, and M13, to the tiny and distant NGC5694. There may be 100 or so other globular clusters that lie hidden behind thick bands of interstellar dust; even though these possible clusters could include the most magnificent cluster in the galaxy, they lie hidden from us. Judging from the numbers of clusters we can see in and near our galaxy, and from the numbers of clusters we see around other galaxies, it is not likely that more than 100 or so lie hidden.

Imagine what our sky would look like if we orbited a star just a few hundred light years from one of the great globulars, and that each night we saw the light of this huge body dominating our sky. How would such a sight affect our perception of the universe? Our religious beliefs? What if we lived inside a globular cluster? With three to 30 stars per cubic light year, stars at the center of a cluster might be separated by just a few light months instead of the four light years to the closest star where we are. The sky would be lit by many first magnitude stars, not a daylight sky, but perhaps a late twilit one.

The classic method of classifying globulars is by their concentration.

Figure 17.2. Globular cluster M79. Tim Hunter of Tucson, Arizona, took this through a 60 cm reflector.

Devised by Harlow Shapley earlier this century, this system evaluates a cluster based on the concentration of the stars as we see them through a telescope. In this system, the Roman numeral I represents the densest and XII the sparsest concentration of stars. Knowing the classification should increase your enjoyment of these objects, I have included it for each globular discussed in this chapter:

>    I: high concentration toward center
>    II: dense central condensation
>    III: strong inner core of stars
>    IV: intermediate rich concentration
>    V: intermediate concentration
>    VI: intermediate
>    VII: intermediate
>    VIII: rather loosely concentrated towards center
>    IX: loose towards center
>    X: loose
>    XI: very loose towards center
>    XII: almost no concentration in center

Discovering new globulars nowadays is a rare event. Because these objects are generally bright, they were all discovered long ago during the golden age of visual observing, when Charles Messier and William and John Herschel were surveying the northern and southern skies. Of course, these early observers did not really understand the nature of what they were looking at, and it is only with the advent of large telescopes and photography that these curious fuzzy patches were being identified as globular clusters. One of the more recent discoveries occurred in 1932, when Clyde Tombaugh found that NGC5694 was a globular cluster.

Where are the globulars in the sky? They seem to be concentrated near the Milky Way, both north and south of the plane of our galaxy. Curiously enough, we haven't seen any right on the plane of our galaxy, at the center of the Milky Way, nor do we see many at great angular distance from the plane. Thus, the sky of winter, offering us views away from the galactic center, has few globular clusters, and in fact M79 is the only good example of one.

## 17.4  Diffuse nebulae

So many objects in the sky seem to be nebulous in appearance that until about 1960 they were all referred to by the Latin term 'nebula' for cloud. Even after the starry nature of the clusters and galaxies was established, the term nebula was so established that removing it was impossible. Nebulae would now be in two classifications: the galactic nebulae, consisting of the genuine gas and dust clouds, and the 'extragalactic nebulae' or outside-our-own-galaxy galaxies. It was only by the 1960s that the term 'galaxy' became totally accepted.

There are two kinds of nebulae, those that shine and those that do not. A cloud of gas and dust that is near a star and shines is known as a bright nebula, and these clouds shine either by reflection of by emission. These wisps of hydrogen gas that are the real nebulae are especially striking under dark sky conditions.

M42, the Great Nebula in Orion, is an *emission nebula* that shines partly due to reflected light but largely because light from the stars of Theta Orionis is absorbed by the gas and then re-emitted, not unlike the process by which a neon sign is lit.

With the thousands of beautiful objects to see in the sky, it is interesting that, when asked for their favorites, most amateurs will choose one of the emission nebulae. Noted for their majestic shapes, these nebulae are very attractive, especially the Orion Nebula M42 and the Eta Carinae Nebula. My own favorite, the Swan Nebula M17, is also a diffuse nebula. Although M42 primarily is greenish, in a 40 cm reflector it shows hues of red and blue; magnificent colors that also are subtle. The more you gaze through a telescope, the clearer and sharper these colors become. Colors are not all; the rest of the beauty lies in the complexity of its shape.

If you have been observing for a while, one thing you surely would know by now is that the sky is subtle; nothing seems obvious. Most of the planetary detail needs aperture, time, and patience to detect, and the colors and shapes I refer to in these nebulae require the same patience. Some emission

Figure 17.3. M42, the Orion Nebula. This 20-minute exposure was taken through a 20 cm Schmidt–Cassegrain telescope, on AGFA 1000 film. Photo by Dan Ward.

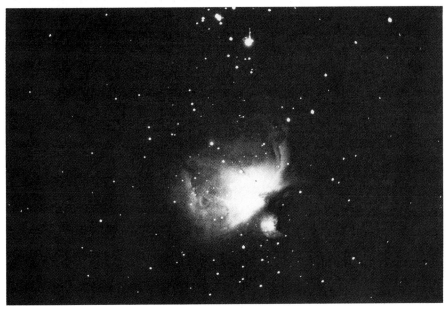

nebulae also glow by reflected light, which further complicates their appearance, adding in a sense another dimension to their beauty.

*Reflection nebulae* shine by light reflected from a nearby star. The Merope Nebula, an example, surrounds one of the stars in the Pleiades, and can be seen with a 15 cm telescope from a very dark sky. A nebula shining in this way shows much less depth of color than would an emission nebula. Also, what color there is would be light blue.

Emission and reflection nebulae are usually not classified like clusters are, as their shapes and sizes are so individual.

Some nebulae shine by emission, others by reflection, and a third variety do not shine at all. The *dark nebulae* are seen only because they block out light of more distant stars. Near the Southern Cross is an area of the Milky Way that resembles a hole punched through the starry sphere; this is actually the Coal Sack Nebula. When John Herschel first observed this and other dark nebulae, he thought that they might be windows to something outside.

Now we know better. The dark nebulae represent one of the most important aspects of stellar astronomy today, for it is in these clouds of unlit gas that the earliest stages of star formation may be taking place.

Figure 17.4. On April 9, 1986, Stephen J. Edberg photographed this beautiful stretch of the southern Milky Way. The star at far left is Alpha Centauri, the closest star system to the Sun, and Beta Centauri is to its right. The dark nebula in the center is the Coal Sack Nebula. The stars to the nebula's upper right form the Southern Cross. The brightness at the right is the Eta Carinae Nebula. 50 mm f/2.8 lens, 10 min exposure, on 3M1000 film.

Since a dark nebula is not lit at all, how do we observe it? Strictly speaking, we can no more see one of these than we can see a black cat in a dark room. We infer the cat's presence by some knowledge of what a cat does; we might even hear or feel it. Similarly, we see a dark nebula largely because of an absence of stars in the midst of a field rich with stars. M42 is primarily a bright nebula, but there is an area just west of Theta Orionis where the brightness drops off abruptly, as do the starry concentrations; there is dark nebulosity there. Even more obvious examples are the Great Rift in the northern Milky Way and, as already mentioned, the Coal Sack in the southern sky.

Since we 'see' dark nebulae only if they are projected against rich starry backgrounds, there must be more of this matter we do not see at all. Early this century, E. E. Barnard made a catalog of all these dark areas that he could photograph; as one of the sharpest observers who ever lived, such a task was very appropriate for him. The Barnard list points to over 100 discrete patches of nebulosity spread over the star-rich Milky Way areas of the sky.

## 17.5  Planetary nebulae

In later stages of the evolution of some stars, an outer layer might be released into a cloud. Depending on the angle at which we see this cloud, the effect might appear as a circle of light, or a ring, or an oval glow, surrounding a faint central star. These clouds are known as planetary nebulae, and the most famous example of one is the Ring Nebula M57. The Ring is so bright that it is visible from almost any site, city or country. Even the smallest department store telescope should show it. It lies directly between the two southernmost stars of the parallelogram of Lyra. In fact, we used to have contests to see how long it took to find M57 in a telescope; I think the record is about four seconds!

One thing you probably won't see is the Ring Nebula's central star. This difficult object is about 15th magnitude, and is surrounded by nebulosity, which makes it even more difficult to detect. I have seen it occasionally through my 40 cm reflector, but it is not easy even with a telescope that large. Not all planetary nebulae have such difficult central stars. The Clownface or Eskimo Nebula in Gemini, NGC2392, has a bright central star, and the Dumbbell Nebula, M27, has a centerpiece that is easy to observe through a 12 inch (30 cm) telescope.

Perhaps the easiest of all central stars is that belonging to NGC6826, known informally as the 'Blinking Planetary'. This sight is hard to believe – if you concentrate on the central star, the nebula then disappears! This is what we might call a 'psycho-visual' effect in which the stars that are surrounded by nebulosity might appear to play games with them; both star and nebula vie for our attention, with the result of our seeing first one, then the other. You should be aware that the strength of this effect depends on the size and focal length of your telescope. Although the Blinking Planetary Nebula is the most famous example of this strange effect, there are others, occurring whenever the central star is close in magnitude to the total magnitude of the entire nebula. Thus, I have seen the Clownface Nebula disappear when I concentrated on its central star too.

Planetary nebulae are classified according to the 'Voroncov–Velyaminov' scheme as follows:

I     = stellar,
IIa   = oval, homogeneously bright, concentrated,
IIb   = same as IIa but not concentrated,
IIIa  = oval, unhomogeneously bright,
IIIb  = same as IIIa but with brighter edges,
IV    = annular,
V     = irregular with intermediate to diffuse nebulosity,
VI    = anomalous, exceptional.

## 17.6  Supernova remnants

While objects like M57 represent small releases of the outer layers of stars, M1, The Crab Nebula, is an example of a far vaster explosion. What we see is a cloud of gas that is the result of a star that has disintegrated. There are two ways that this can happen; in a double star, one member might continually transfer matter to another until a physical limit is reached, after which the second star completely destroys itself. The result is a type I supernova. The other mechanism involves a star nuclearly 'burning' elements into heavier and heavier elements, resulting in still heavier elements until finally it tries to burn iron. When the attempt to 'burn' the iron fails, the core of the star collapses, resulting in the gigantic explosion of a type II supernova.

In a type II supernova, all that is left of the star is a small, rotating 'neutron star' surrounded by a cloud of material. Sometimes we get to see this cloud. There are very few examples of supernova remnants that we can see; besides the Crab, Cygnus offers a huge structure called the Veil Nebula, and in the southern hemisphere the Gum Nebula occupies several square degrees of sky.

## 17.7  Galaxies

Of all the objects we see in the sky, the farthest are the galaxies. Past the star clusters and nebulae, beyond the farthest reaches of our own Milky Way, are the other galaxies. Formerly called the 'extragalactic nebulae' and 'island universes', these distant objects are simply galaxies. If you are reading this in March or April, and if you have access to a dark sky, you can see

Figure 17.5. This stunning view of the Whirlpool Galaxy, M51, was taken by Tim Hunter through his 60 cm reflector, in a 10-minute exposure.

Figure 17.6. Small and Large Magellanic Clouds, taken in a four-minute exposure by Dan Ward. Konica 3200, 28 mm lens at f/2.8.

more galaxies than you know what to do with. Draw an imaginary line between Arcturus and Denebola, the tail of Leo, the Lion. Then point your telescope to a place about the middle of that line and slowly move the telescope south. You won't have too many fields of view to check before you see the first of many galaxies of different shapes and sizes. Some will appear narrow and elongated, and spiral-shaped in larger telescopes. A few will look like elliptical balls, and very few will resemble very faint, barely visible hazes.

Galaxies look like many things, depending on how large they are, how they rotate about their centers, how much dark matter they have, and at what angle they present themselves to us. Using the 100 and 200 inch telescopes in California, Edwin Hubble classified galaxies into several broad types. The classification reads thus:

E  = elliptical, subdivided 0 to 7,
S  = spiral,
SB = barred spiral,
I  = irregular,

For spirals:

a = bright nucleus, arms less developed,
b = lesser nucleus, arms well developed,
c = weak nucleus, arms very conspicuous,
p = peculiar.

Our own Milky Way is a *spiral galaxy*; surrounding an oval center are huge arms of stars that make the whole array resemble a pinwheel. Our spiral galaxy rotates about its center every 220 million years. Think of the dinosaurs of the Triassic period roaming around about one galactic year ago and you will have an idea of what that time scale means.

Spiral galaxies can show special features; M64 has a huge blot of dark matter near its center, making its appearance so striking that we call it the 'Black Eye Galaxy'. More commonly, galaxies may have thin lanes of dust that cross their diameters. A 20 cm reflector under a good sky should reveal such a lane in the Andromeda Galaxy.

As some spirals rotate, they develop linear extensions, so that the arms do not begin at the centers but at the ends of the extensions. These galaxies are called 'barred spirals'.

*Irregular* galaxies include a variety of galaxies of different appearances, from the Small Magellanic Cloud, a neighbor galaxy to the Milky Way, to M82 in Ursa Major. At first cosmologists thought that the galaxy was undergoing some kind of massive explosion, perhaps tearing itself apart; at least there is evidence of star formation occurring on a huge scale. While many spiral galaxies have dark lanes crossing their lengths, M82 is unusual in having a dark lane crossing its width.

*Elliptical* galaxies are large masses of stars that look like what they are called. M87 is perhaps the best known of the elliptical galaxies, and it is also one of the biggest, containing about five trillion ($5 \times 10^{12}$) stars. Lacking much interstellar dust, the ellipticals tend to have the simplest appearances.

This little guide to the galaxies should be sufficient to get you started. Even though they are very distant, these massive swarms of stars and other matter are often easy to classify, even through a small telescope.

## 17.8  Quasars

One way of considering these star-like, or 'quasi-stellar' objects, is as a type of galaxy. Most astronomers think that the quasars are very far away, billions of light years distant, and that they are extremely energetic centers of galaxies whose more normally luminous outer parts are invisible to us. 3C273 in Virgo, at 12th magnitude, is one of the brightest known examples. With any amateur telescope this object simply looks like a star.

## 17.9  Telescope and sky

What kind of telescope do you need for deep sky observing? In brief, the bigger the better. Usually a 20 cm telescope is recommended for good deep sky views. But what is just as important is the quality of the sky through which that telescope must look. To do justice to these distant patches of light, you need the darkest sky you can find. These two requirements of sky and telescope are quite distinct, and one should not be exchanged for the other. For instance, you may have heard something to the effect that 'a 30 cm telescope in the city is equivalent to a 15 cm in the country'. This is nonsense. For almost every galaxy, and for most of the clusters, a 15 cm reflector out under a dark country sky gives a much more satisfying view, with greater contrast between object and background, than would a 30 cm under the hazy, polluted sky of a city. There is no trade-off; it is simply better to view deep sky objects with any telescope in a dark sky. The two requirements are separate and distinct.

Remember that magnitudes that are published for deep sky objects may be deceptive because they do not take into account surface brightness; a small, bright galaxy easily seen from a city sky may have the same listed magnitude as a large and faint galaxy that is visible only from a dark sky. For faint objects, try averted vision to bring out their details. Finally, the use of a nebular filter (also known as ultra high contrast or light pollution reduction filter) may bring out details of an emission nebula like M42.

## 17.10 For a city sky

Even if you are unable to leave your urban observing site, you still could try observing some of the objects of the deep sky. Here are a few that will reward a city-bound observer:

### January–March

| | |
|---|---|
| M35 | Gemini. Open cluster. Class e. |
| M36, 37, 38 | Auriga. Interesting open clusters, classes f, f, and e. |
| 47 Tucanae | The southern hemisphere's summer spectacular; acknowledged by almost all who have seen it to be the best of the globulars. In addition to being relatively near, this cluster is also intrinsically huge. Class III. |
| M42 | Orion. The Great Nebula. Even from the city this cloud of nebulosity, laced with young blue stars, can be a stunning sight. |

### April–June

| | |
|---|---|
| M81, 82 | Ursa Major. Fine pair of galaxies. M81 is type Sb. M82 has a long, mottled, tortured look, giving an impression of mystery that is appropriate for this unique galaxy. Type I. |
| M94 | Canes Venatici. One of the highest surface brightnesses of any galaxy, this object is easily seen with a 20 cm or larger telescope, even under a sub-urban sky. |
| M13 | Hercules. Finest globular cluster in the northern sky. Shapley class V. |
| M3 | Canes Venatici. Another fine globular. Class VI. |
| Omega Centauri | Centaurus. Also a southern hemisphere globular, where of course it is an autumn object, but far enough north that observers in the far southern United States can see it in their spring. This unique globular cluster is not absolutely circular but somewhat oval, indicating that it rotates. Class VIII. |
| M44 | Cancer. Class d. |

### July–September

| | |
|---|---|
| M4 | Scorpius. One of the nearest of the globulars, and also one of the least concentrated. Class IX. |
| M6, M7 | Scorpius. Two fine open clusters, easily found and enjoyed. Both are class e. |
| M11 | Scutum. Very rich open cluster; lots of blue stars. Class g. |
| M22 | Sagittarius. If this relatively nearby globular were not in the middle of a rich Milky Way star field, it would look far more dramatic. Class VII. |
| M39 | Cygnus. Fine open cluster. Class e. |
| M57 | Lyra. The Ring Nebula, a truly fabulous planetary nebula. Class IV. |

Figure 17.7. Omega Centauri, one of the most prominent globular clusters. NGC5128, a bright galaxy, is the distinctly unstar-like blob of light to the lower left. South is up in this photograph, taken by Steve Edberg on May 30, 1987, with a 260 mm f/5.2 lens, 10-minute exposure, on 3M1000 film.

**October–December**

| | |
|---|---|
| M15 | Pegasus. Exquisite globular cluster. Class IV. |
| M31 | Andromeda. Even though this is a magnificent spiral galaxy, only its central portion is visible from city sites; to see its complex outer arms you really need a dark location. Use an eyepiece that gives as wide a field of view as possible. Type Sb. Its companion M32 is easily visible from a city, although M110, its other companion with lower surface brightness, is not. |
| M45 | Taurus. The Pleiades. Class c. |
| The Hyades | Taurus. One of the closest open clusters, some 130 light years away. Because it is so close, the motions of its members relative to one another can be determined. Class c. |
| NGC869 and 884 | Perseus. The Double Cluster. Exquisite. Class f and e. |

## 17.11 For a dark sky

Under a country sky, all the previously listed objects, which teased your fancy, will now unravel their veils of secrecy and emerge in full splendor. And there is more! A brief sampling:

**January–March**

| | |
|---|---|
| M78 | Orion. A nebula whose shape resembles that of a faint comet. A bright reflection nebula. |
| M79 | Lepus. Like most of us, globulars do not like winter. If it is mid-January and you can't wait for a globular cluster fix, try this one south of Orion. Class V. |
| NGC2237 | Monoceros. The Rosette Nebula and its associated open cluster NGC2244 are a beautiful sight, especially through 20 cm or larger telescopes. Views of nebulae like the Rosette are especially enhanced if you attach a nebular filter to your eyepiece. |
| NGC2264 | Monoceros. An open cluster near the dark 'Cone Nebula'. |

**April–June**

| | |
|---|---|
| M51 | Canes Venatici. The Whirlpool Galaxy. Type Sc. |
| M87 | Virgo. Elliptical galaxy. Hard to find among its numerous neighbor galaxies. A huge mass of several trillion stars. Type E0. |

M84 and M86    Virgo. The brightest of a truly inspiring field of telescope view. How many galaxies can you see in that
eyepiece field? These galaxies, with M87, are part of
the Coma–Virgo 'supercluster' of galaxies. Types E1
and E3.

Figure 17.8. Elliptical galaxy M87. Photo by Tim Hunter. 60 cm f/5 reflector.

Figure 17.9. NGC6946, a ninth magnitude galaxy in Cepheus. Sketched by Dan
Ward using a 42 cm f/4.5 reflector.

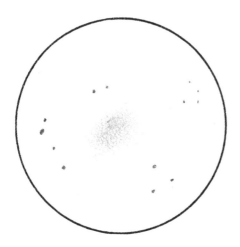

| NGC4565 | Coma Berenices. A fine spiral galaxy viewed edge-on. Type Sb. |

**July–September**

| M8 | Sagittarius. The large Lagoon Nebula. |
| M20 | Sagittarius. The intriguing Trifid Nebula. |
| M17 | Sagittarius. The Swan or Omega Nebula. Of all the Messier objects, the strange wisps and subtle curls of nebulosity of the Omega make this my favorite. The open cluster associated with it is class c. |
| NGC6826 | Cygnus. The Blinking Planetary Nebula. Experiment with different eyepieces. Concentrate alternately on the nebula and then on its relatively bright central star. When you look at the star the surrounding nebula might disappear. Class IIIa. |

**October–December**

| M1 | Taurus. The Crab Nebula. Remains of the supernova of 1054. |
| M33 | Triangulum. Needs a dark sky and wide field of view for this nearby galaxy. A difficult but spectacular example of a galaxy viewed face-on. Type Sc. |
| NGC253 | Sculptor. A beautiful, almost edge-on galaxy, but far south. Reminiscent of M31. Type Scp (the p for peculiar). |

This list is very selective. It could go on and on, but I hope that once you have given the city and country lists a chance, you'll be hooked forever on this tantalizing field.

Unless you happen to discover a supernova in one of the galaxies, visually observing deep sky objects through a small telescope has little professional value. Deep sky is mainly for fun, but because there is so much up there, and because each telescope offers a different interpretation of each object, and because each eyepiece changes what you see in each telescope, each view will be unique and the fun should never end.

# 18  Messier hunting

Why would you want to find all of the Messier objects? After all, they represent a part of astronomy's past, the disappointments of a late eighteenth century comet hunter. We have come so far since then!

Messier's hunt for comets was rather successful, but the dozen comets that bear his name have all gone. What have not gone are the hundred plus objects that caused him a thrill and then disappointment as, one by one, he listed them in his catalog of objects to avoid.

We observe the Messier objects because they can teach us so much. These 110 objects are scattered around the sky, offering special treats for any season. They represent almost every kind of 'faint fuzzy' – open and globular star clusters, bright nebulae, reflection nebulae, planetary nebulae, galaxies, even (M40) just two faint stars.

Although Messier hunting is traditionally a 'lone wolf' sport, it is more fun at least to keep score in a group, like a friendly competition. In the 1940s, the RASC's Montreal Centre formed what is probably the first formally organized Messier club. Its goal was and still is a friendly competition to find all the objects observed and recorded by Charles Messier. As part of the observing programs of astronomical societies, Messier clubs have sprung up all around the world to encourage people to search for, find, draw, and photograph these objects.

The hunt for these objects Messier wanted to avoid is fabulous training both in recognizing constellations and in identifying the different types of deep sky objects. Since the Messiers are placed all over the northern sky they can teach you the constellations in a way you are not likely to forget. Because catalog objects range from 4th to 12th magnitude they offer objects for us to observe with telescopes of all sizes. Their variety of types, from clusters to nebulae to galaxies, tell you how to become familiar with these

Figure 18.1. The Andromeda Galaxy M31, captured in a 25-minute exposure through a 60 cm f/5 reflector by Tim Hunter.

214 The Sky: A User's Guide

types of objects by looking at their best representatives. Also, their variety of appearances alert you to the differences between diffuse and condensed objects.

For this work you should use at least a 10 cm telescope, and to see them all, at least a 15 cm instrument. Binoculars are quite useful in catching some Messiers (some observers have seen almost all the Messier objects that way), but because you really want to enjoy the details of each Messier, I recommend a telescope for all of them. For each Messier, record carefully the date, the place, sky conditions, instrument used, and the eyepiece, and then comment on or sketch the details of each one you see.

It took me five years – from September, 1962, to May, 1967 – to spot all the Messiers. The first I saw was M31, the Andromeda Galaxy. Using a small 8 cm reflector, I carefully found a star named Mirach, or Beta Andromedae, and moved north past a pattern of stars that matched that in a book I was using. The most difficult problem was to figure out the scale: how much of the sky was I seeing in each field of the eyepiece? At first I had no idea and could not find any of the stars, but suddenly the entire pattern fell into place and on I moved to a system of three stars, one bright, two faint, acting almost as pointers. Moving a little more than a field away, I expected to see more and more fields of just plain stars. Then suddenly the field was lit by a huge white cloud, quite faint, that could only be the huge galaxy for which I was searching! I was astonished. To actually set out on a search for an object that couldn't be seen with the naked eye from my city-bound observing site, and then to find it suddenly appearing in a telescope, was a special experience.

Quickly I found that the Messier competition was a sporting way to enter into the mysterious and distant realm of the clusters and galaxies. As I moved through the list, I saw the brighter ones first and saved the complex region of galaxies in the Coma Berenices and Virgo region until the end. The hardest to find in the early years was M1, the Crab Nebula, which is very diffuse and almost impossible to see through a small 8 cm telescope from a city sky. I remember making attempts on nine different nights to find this nebula, thinking that no. 1 in anybody's catalog must be very bright and easy to see. Finally, with a larger, 12.5 cm telescope, on September 1, 1963, I saw a faint smudge of light that was the remains of a star that exploded 909 years, 1 month and 28 days earlier.

The Messier list that follows is presented in two ways. The first order was designed originally by Alan Dyer, a prominent deep sky observer from Edmonton, Alberta, and published in the RASC's *Observer's Handbook*. Instead of being arranged numerically, it is arranged by season for the convenience of an observer watching in the evening hours of different parts of the year. Next I will offer the list in a different order, better suited for trying to see them all in a single night.

**December–February sky**

M1    Taurus    The Crab Nebula. Remnant of the supernova of July 4, 1054. Deceptively simple – actually, if you are not in a dark country sky, you may not see it since the sky brightness may be higher than that of the object.

M45    Taurus    The Pleiades. One of the best known and most easily seen of the Messiers. One of the stars, Merope, is surrounded by a nebula that shines by light reflected from the star.

M36    Auriga    Open cluster, covering a wide area. This beautiful cluster features stars in apparent 'lanes'.

M37    Auriga    Open cluster, some observers call it the nicest of the Auriga clusters. Comprises a mass of faint stars.

M38    Auriga    My personal favorite of the Auriga open clusters. A 10 cm telescope shows many stars in this cluster.

M42    Orion    The Great Nebula in Orion. This has got to be one of the most stunning objects in the entire sky. It is visible faintly with the unaided eye under a dark sky, but binoculars will reveal its presence even from downtown. Any telescope will show this object well. It consists of a group of young, hot, blue stars, many of which vary in brightness (although you will not notice this variation through casual observation), embedded in bright clouds of nebulosity. Near the center of the whole scene lie the four stars of Theta-1 Orionis, known as the Trapezium.

Figure 18.2. M45, the Pleiades star cluster. Photo by Tim Hunter.

M43   Orion   Actually just an extension of M42.

M78   Orion   A reflection nebula that is interesting because it resembles a telescopic comet.

M79   Lepus   A rare example of a globular cluster in a part of the sky far removed from the center of the Milky Way.

M35   Gemini   Open cluster; one of the prettiest anywhere. Visible faintly through binoculars, myriads of stars seen through any telescope.

M41   Canis Major   A 'spread out' open cluster just south of Sirius. Easy to see even from a city sky.

M50   Monoceros   Fine open cluster, lots of faint stars.

M46   Puppis   This beautiful open cluster includes a planetary nebula, NGC2438.

M47   Puppis   Open cluster; use low power for best view.

M93   Puppis   A small, bright open cluster.

M48   Hydra   Bright open cluster.

## March–May sky

M44   Cancer   The Beehive open cluster, also known as the Praesepe. Several groups of stars in this cluster appear as small triangles. Visible as a diffuse object with the naked eye from a dark sky; in fact it does resemble a big, diffuse comet when it is low in the sky. The poet Gerard Manley Hopkins (see Chapter 23) saw it that way late in the 19th century. Coincidentally, a big, diffuse, naked eye comet (IRAS–Araki–Alcock) actually passed a few degrees from this cluster in the spring of 1983.

M67   Cancer   Much smaller and less impressive open cluster than M44. One of the oldest open clusters known.

M40   Ursa Major   Two stars close together, that Messier might have interpreted as diffuse.

M81   Ursa Major   Beautiful spiral galaxy, although the arms might not be detectable in a 10 cm telescope.

M82   Ursa Major   Peculiar galaxy, has a long, thin appearance. Through a 20 cm or larger telescope, dark lanes cross it irregularly. One of the most controversial deep sky objects, astronomers once thought that it was exploding. Later fashion suggested some major disruption taking place at the core. Current thinking is that a burst of star formation is taking place in M82.

M97   Ursa Major   The Owl Nebula, a large, faint, diffuse planetary nebula. Like the Crab, this nebula will get washed out under a bright city sky, but unlike the Crab, this nebula is only the remains of a relatively small disruption in the atmosphere of its central star. One of the more diffuse of the planetary nebulae, this nebula is not easy under a city's light polluted sky.

M101   Ursa Major   Beautiful spiral galaxy with wide, open arms and lots of small concentrations of hydrogen called H II (pronounced H two) regions. Although this galaxy is faint, it is quite complex and is an excellent choice for study under a dark sky with a 20 cm or larger telescope.

M108  Ursa Major  An almost edge-on spiral galaxy not far from the Owl Nebula, M97. Relatively condensed, visible easily through a 10 cm telescope under good conditions.

M109  Ursa Major  A barred spiral galaxy, although the 'bar' is not seen through a 10 cm telescope. In this type of galaxy the arms do not spiral out directly from the center, but instead from the bar.

M65  Leo  A very nice spiral galaxy.

M66  Leo  In the same field as M65. Both these galaxies are bright enough to be observed from a suburban location through a 15 cm or larger telescope.

M95  Leo  One of the best examples of a barred spiral galaxy.

M96  Leo  Spiral galaxy in same low power field as M95.

M105  Leo  Elliptical galaxy close to M95 and 96.

M53  Coma Berenices  Globular cluster; stars can be resolved through a 15 cm telescope.

M64  Coma Berenices  Spiral galaxy with large, oval dust lane that gives it the name 'Black Eye Galaxy'. You need a 15 cm telescope to see the dust lane clearly.

M85  Coma Berenices  Elliptical galaxy, a bright smudge through a 15 cm telescope.

M88  Coma Berenices  Spiral galaxy with many arms. With a 20 cm telescope, the complex structure of the arms becomes apparent.

M91  Coma Berenices  Barred spiral galaxy, although the bar is not apparent through a 10 cm telescope.

M98  Coma Berenices  Essentially an edge-on spiral galaxy.

M99  Coma Berenices  Spiral galaxy; looks oval through a small telescope.

M100  Coma Berenices  Spiral galaxy with bright nucleus. The bright center is striking in a 15 cm telescope.

M49  Virgo  Elliptical galaxy, seen as a misty spot through a 10 cm telescope. Oval shape is more apparent through a 15 cm instrument.

M58  Virgo  Barred spiral galaxy; bar can be seen through a 20 cm telescope.

M59  Virgo  Elliptical galaxy; small and quite bright, even through a 10 cm telescope.

M60  Virgo  Elliptical galaxy. M58, 59, and 60 make an interesting trio of galaxies; with a 20 cm telescope you can differentiate between the spiral type of M58 and the elliptical shapes of the others.

M61    Virgo    Spiral galaxy with low surface brightness; one of the most difficult of the Messiers to spot.

M84    Virgo    Elliptical galaxy; small and compact through a 15 cm telescope.

M86    Virgo    Elliptical galaxy. Shares field with M84 and several fainter galaxies. The one-degree field looks almost like the constellation of Sagitta the arrow, but each 'star' is a galaxy!

M87    Virgo    Gigantic elliptical galaxy, probably contains more than five trillion stars. Images with large telescopes show a thin, straight jet of light coming from the nucleus. I have seen this jet visually with a 154 cm (61 inch) reflector, but not with a 40 cm.

M89    Virgo    Elliptical galaxy, smaller than M87.

M90    Virgo    Spiral galaxy near M89.

M104    Virgo    The Sombrero Galaxy. It really does resemble a hat!

M3    Canes Venatici    Marvellous globular cluster.

M51    Canes Venatici    The Whirlpool Galaxy, contains numerous arms, one of which leads to a companion galaxy. Beautiful in a 15 cm telescope.

M63    Canes Venatici    Spiral galaxy known as the Sunflower.

M94    Canes Venatici    Spiral galaxy resembling a comet through a 15 cm telescope.

M106    Canes Venatici    Spiral galaxy.

M68    Hydra    Faint globular cluster.

M83    Hydra    Very exciting spiral galaxy, with complex spiral structure visible through 20 cm or larger telescopes. Use as much aperture as possible to bring out details of this structure.

M102    Draco    Elliptical galaxy. There is uncertainty as to whether Messier really meant this object, or if he had seen something else instead.

M5    Serpens    Bright, large globular cluster. Very nice even through a 10 cm telescope.

## June–August sky

M13    Hercules    What a marvellous globular cluster this is! Use as high a power as the sky conditions of seeing will permit. Some nineteenth century observers reported some dark lanes through this cluster that do not always appear on modern photographs. However, with a 40 cm telescope, the Canadian observer Alister Ling reported seeing these same lanes.

M92    Hercules    Bright globular cluster. Begins to resolve into stars with 60 power.

M9    Ophiuchus    Globular cluster with a bright core. In the same field of a low power eyepiece, you should also see NGC6356.

M10    Ophiuchus    Globular cluster, strikingly rich.

M12    Ophiuchus    Globular cluster, not nearly as rich as M10.

M14   Ophiuchus   Globular cluster, compact in appearance.

M19   Ophiuchus   Globular cluster; one of the few that is not circular in appearance (Omega Centauri is another).

M62   Ophiuchus   One of the brightest globular clusters, quite compact in appearance.

M107   Ophiuchus   Globular cluster; difficult to resolve into stars with telescopes smaller than 10 cm.

M4   Scorpius   Walter Scott Houston, one of this century's best known deep sky observers, once said that 'there is not just one M4; there are many.' Try this globular through a number of different telescopes; each will bring out different details.

M6   Scorpius   Magnificent open cluster, easily visible to the naked eye under a dark sky.

M7   Scorpius   Also a magnificent open cluster, seen with naked eye under dark conditions.

M80   Scorpius   Small, hard to resolve globular cluster.

M16   Serpens   An emission nebula, shining because radiation from nearby stars excites atoms within the nebula, causing them to glow. This is the 'Star-Queen' nebula, and it includes an open cluster.

M8   Sagittarius   The Lagoon emission nebula. Very complex region where, like in other emission nebulae, stars are forming or are about to form. Large telescopes reveal tiny dark areas called 'Bok Globules', which are named for the astronomer Bart Bok who first theorized that they are areas where star formation is beginning to take place. Beautiful sight even through a 10 cm telescope.

M17   Sagittarius   The Swan Nebula; also known as the Omega Nebula because part of it resembles the Greek letter. This is my preference for the most beautiful Messier, nicer even than M42. Its subtlety is most impressive, especially through a 30 cm or larger telescope. In addition, it lies in a marvellous field of Milky Way stars.

M18   Sagittarius   This open cluster lies just one degree south of M17.

M20   Sagittarius   The Trifid emission nebula. Dark lanes appear to divide this complex into three parts, well seen in a 20 cm telescope.

M21   Sagittarius   Open cluster less than a degree northeast of M20.

M22   Sagittarius   This globular cluster rivals M13 in size and beauty, although northern hemisphere dwellers may not appreciate this since the cluster is so far south, and because it is in a rich field in the Milky Way.

M23   Sagittarius   Open cluster; comprises bright stars and surrounded by a rich field of Milky Way stars.

M24    Sagittarius    Not a specific cluster, this is just a bright star cloud of the Milky Way.

M25    Sagittarius    Rich open cluster in the midst of a rich field of stars.

M28    Sagittarius    Globular cluster; bright and easy to resolve in a 10 cm telescope.

M54    Sagittarius    Small globular cluster; difficult to resolve even through 20 cm telescopes.

M55    Sagittarius    Large, easily resolved globular cluster.

M69    Sagittarius    Relatively faint globular cluster; difficult to resolve. Nearby, in the same low power field of view, is NGC6652, another, fainter, globular cluster.

M70    Sagittarius    Globular cluster, faint in a 10 cm telescope.

M75    Sagittarius    One of the most distant of the Messier globulars.

M11    Scutum    The Wild Duck. This open cluster is made very striking by the presence of a bright foreground star.

M26    Scutum    Open cluster, slightly fan-shaped.

M56    Lyra    Globular cluster. Although a 10 cm telescope might show this cluster as cometary in appearance, a larger telescope should begin to show a mottled look.

M57    Lyra    The best known planetary nebula. This bright, sharp object appears nicely in any telescope. I saw the central star through a 40 cm telescope; although it is about 15th magnitude, because the middle of M57 is nebulous, the star is harder to see than other stars because of its brightness. Observers have reported varying degrees of difficulty detecting the star.

M71 Sagitta    Globular cluster so loose it could be confused for an open cluster. Rich field of surrounding stars.

M27 Vulpecula    The Dumbbell Nebula. This planetary nebula actually resembles a dumbbell, with two edges larger than the center. Its central star is not difficult to see through a 30 cm or larger telescope. Like other planetary nebulae, M27 has a relatively high surface brightness compared to other deep sky objects of its diameter, and is not difficult to spot from a suburban sky.

M29 Cygnus    Open cluster, difficult to detect because it lies in such a rich Milky Way field.

M39 Cygnus    Bright by loosely defined open cluster.

## September–November sky

M2    Aquarius    Globular cluster with bright core surrounded by dimmer halo of unresolved stars.

M72    Aquarius    Smaller globular cluster; faint and not resolved into stars with most small telescopes.

M73   Aquarius   Nothing really, just a forlorn-looking group of four stars.

M15   Pegasus   My favorite fall globular cluster. Bright and compact. A 15 cm telescope should start to resolve the stars in the halo of this cluster.

M30   Capricornus   Oval shaped globular cluster; bright center.

M52   Cassiopeia   Bright, rich open cluster.

M103   Cassiopeia   Open cluster, difficult to find since it is in the middle of a rich Milky Way field.

M31   Andromeda   The Great Galaxy of Andromeda. Under a dark sky this galaxy is clearly visible without optical aid, and binoculars show it extending over several degrees of sky. The more telescope you use on this object, the more detail you will see. A 20 cm, for example, will show the dark dust lane that crosses the galaxy's length, and larger instruments should reveal the compact bright spots that are H II (hydrogen) regions.

M32   Andromeda   Bright, compact, elliptical galaxy, a companion to M31. It looks like a fuzzy star.

M110   Andromeda   Until recently, this elliptical galaxy was known simply by its NGC number, 205. Evidence suggesting that Messier had recorded this object has led to its addition to the catalog. Visible as a misty spot through a 10 cm reflector.

M33   Triangulum   A relatively close spiral galaxy, but it is large and diffuse. A city sky is brighter than the surface brightness of M33, rendering it invisible unless you have sufficiently dark conditions.

M74   Pisces   One of the hardest of the Messiers to see, although some observers with 30 cm or larger telescopes have called it a faint 'cousin' of M51 in appearance.

M77   Cetus   Spiral galaxy. One of the Seyfert galaxies that are typified by bright, star-like cores. That effect is seen through at least a 15 cm telescope.

M34   Perseus   Open cluster, covers wide area.

M76   Perseus   The Little Dumbbell planetary nebula; through a 20 cm it really begins to resemble its larger namesake, M27.

# 18.1   Messier marathons

Some clubs now offer 'marathon nights' for observers to try to see how many Messiers their members can spot in a single night. If you hold your marathon during the middle of March, there is a chance to spot all but one of the objects in one night. On March 15, 1983, I did just that, using the opportunity to reacquaint myself with all of them through the point of view of my (40 cm) telescope. I did not find all of them; M30 rose over a distant southeast mountainside after the sky was too bright. This session was a pleasure, and I was stunned by the variety of remote objects that cling to the sky, at their great distances.

Although Messier marathons are fun, I do not suggest that you try one simply to record lots of objects. Enjoy each one; examine the more interesting Messiers at different powers. Make your night a quality one. During my own session, I found time to make some rough notes which I reproduce below. I also had time to see some NGC objects, and even warm up inside from time to time.

Messier Marathon March 15/16, 1983 Session 6207AN
No. Messiers seen = 109
No. other NGCs seen (numbered and identified) = 34
(Order of objects courtesy of Don Machholz and John Kerns, *Deep Sky Monthly* March 1982 issue, page 8.)

| Messier | Time | |
|---|---|---|
| 33 | 1931 | Very difficult in twilight. |
| 77 | 1933 | Compact and concentrated. |
| 74 | 1943 | Very diffuse. |
| 31 | 1949 | No dust lane visible, low in North, twilight. |
| 32 | 1949 | Dwarf elliptical all right . . . obviously shaped like one! |
| 110 | 1949 | Quite a galaxy in its own right! |
| 76 | 1953 | Really does look like a miniature M27! Its two parts are more separated than in M27. |
| 34 | 1956 | Extremely loose open cluster. |
| 45 | | Merope nebula easily visible both in finder and in main scope. |
| 79 | 2034 | With 20 cm, since it was too far south to be seen with 40 cm. |
| 42 | 2036 | Breathtaking. |
| 43 | 2036 | |
| 78 | 2037 | Visible with its companion, and also a brightening in the sky at the position of Barnard's loop. |
| 41 | 2038 | Nice open cluster. |
| 93 | 2044 | Beautiful open cluster. NGC2482 nearby. |
| 47 | 2045 | Coarser than M46. |
| 46 | 2046 | Unique. Planetary nebula inside, NGC2438. |
| 50 | 2050 | Intermediate rich open cluster. |
| 48 | 2052 | Coarser than M50. |
| 1 | 2055 | Crab-like (!) Ethereal. |
| 35 | 2056 | Saw also faint open cluster nearby. |
| 38 | 2057 | Blue, faint stars. My favorite of the Auriga open clusters. |
| 36 | 2058 | Saw also planetary nebula NGC1931 nearby. |
| 37 | 2059 | Saw also open cluster NGC1907 nearby. |
| 44 | 2102 | Many stars apparently arranged in triangles. |
| 67 | 2105 | Here, some stars arranged as a question mark. Fine open cluster. |
| 95 | 2110 | Of next 3 galaxies, covers most area. |
| 96 | 2110 | Most distinctly visible of 3. |
| 105 | 2110 | Smallest but brightest of three. Also saw two neighbors, NGC3384, 3389. |

| | | |
|---|---|---|
| 65 | 2113 | Brighter than M66. |
| 66 | 2114 | Longer than M65. Also saw NGC3628 and faint NGC3593. |
| 81 | 2116 | Exquisite. |
| 82 | 2116 | Dust lane does appear to have been 'turned' toward M81. NGC3077 and 2976 also; 2976 larger of two. |
| 40 | 2121 | Double star. Also saw NGC4290, 20' to west. |
| 108 | 2126 | Long spiral. |
| 97 | 2126 | Beautiful, soft, round nebula. |
| 109 | 2130 | Almost like a very distant M31. Similar aspect. |
| 106 | 2134 | Hint of dust lane. Tiny NGC4217 close by. |
| 94 | 2135 | Very bright. |
| 63 | 2138 | Comet-like, in a sense. Bright, seen also through 7.5 cm finder. |
| 51 | 2140 | In reverence, I look at this magnificent sight. |
| 101 | 2142 | Spreads out. |
| 102 | 2147 | Identity of this one is historically uncertain. NGC5856 is small, NGC5907 is long, NGC5879 faint and small. |
| 3 | 2151 | Beautiful globular cluster. (This whole night is like a reunion of old friends!) |
| 64 | 2156 | Dust 'lane' clear using 16.3 mm eyepiece. |
| 53 | 2159 | Brilliant. Nearby, NGC5053 much fainter but larger. |
| 85 | 2205 | Seen with NGC4293 and 4294. |
| 100 | 2208 | Difficult to identify because of crowded field. I recall observing a supernova in M100 in 1979. |
| 99 | 2212 | Arms visible nicely at 160×. |
| 98 | 2222 | Long spiral. |
| | | NGC4565. Pity Messier missed this immense spiral galaxy. |
| 84 | 2229 | Saw 8 galaxies in the same low power (53×) field. Two of them were identified as NGC4435 and 4438. |
| 86 | 2229 | |
| 90 | 2229 | Delicate. |
| | | Look up from the telescope for a short while to look at the sky; Gegenschein very noticeable. |
| 89 | 2231 | Small elliptical galaxy near M90. |
| 87 | 2235 | Impressive, especially at 160×. Jet not visible! |
| 88 | 2246 | Beautiful spiral. Also confirmed NGC4595, 4571, 4540. |
| 91 | 2247 | Fainter and rounder than M88. (=NGC4548; barred spiral shape not seen clearly.) |
| 58 | 2310 | Can see barred spiral shape with a little imagination, high power. Spectacular 'colliding' (apparently) galaxies NGC4567, 4568 nearby. |
| 59 | 2320 | NGC4638 nearby, also M60. |
| 60 | 2320 | NGC4647 almost touches it. |
| 49 | 2330 | Bright. |
| 61 | 2332 | Faint but large 'Sc'–type spiral. Has wide open arms. |
| 104 | 2340 | Magnificent! Dust lane clear at high power. |
| 68 | 2343 | Perfectly resolvable at high power. |
| 83 | 2349 | Bright center (or nearby superimposed star). Ethereal; beautiful object. |

| 5 | 2356 | Resolved even using 10 cm. |
|---|---|---|
| 13 | 0045 | Magnificent at 160×. Nearby to north is NGC6207, a galaxy seen easily as well. |
| 92 | 0050 | Coarser than M13. Also very beautiful. |
| | 0055 | Omega Centauri with 10 cm. Easily resolved. |
| 14 | 0227 | A very fine, hard to resolve globular cluster. |
| 10 | 0230 | Coarser than M14: easy to resolve. |
| 12 | 0232 | A fine, easy to resolve globular cluster between M10 and M14 in richness. |
| 107 | 0235 | Much smaller in size than M10, 12, or 14. |
| 9 | 0240 | Easily resolvable. Seen also in 10 cm. |
| 80 | 0243 | Small, bright. Didn't see this one in 10 cm. |
| 4 | 0244 | In 40 cm, M4 is stunning, absolutely stunning. |
| 19 | 0247 | A tight globular cluster in a very rich field. |
| 62 | 0250 | Tight. N. side of 'halo' slightly larger than S. |
| 57 | 0308 | A 'confirmed glimpse' of the central star, using the 12 mm and 6 mm eyepieces. |
| 56 | 0314 | Faint, small, and fairly coarse. |
| 27 | 0317 | Central star very prominent at 160×. |
| 71 | 0319 | Unusual shape for globular cluster. |
| 11 | 0320 | Exquisite. |
| 7 | 0325 | Large open cluster; stunning naked eye object. |
| 6 | 0326 | Large open cluster. |
| 26 | 0334 | Small open cluster. |
| 16 | 0335 | Incredible open cluster with nebulosity. |
| 17 | 0338 | Running out of words to describe these magnificent objects. Nebular filter provides one of the finest sights – earth or sky – that I have ever seen. |
| 18 | 0346 | Loose, small open cluster. |
| 24 | 0350 | Huge area with NGC6603. |
| 25 | 0356 | Blue stars in this open cluster. |
| 23 | 0404 | Nice open cluster. Planetary nebula NGC6445 and globular cluster NGC seen nearby. |
| 21 | 0410 | Weak open cluster near M20, the Trifid. |
| 20 | 0411 | Trifid Nebula, best with nebular filter. |
| 8 | 0412 | Nebular filter the way to go with this stunning sight. |
| 22 | 0415 | My God! Fantastic! |
| 28 | 0417 | Bright, small, extremely rich field. |
| 29 | 0422 | Weak open cluster. |
| 39 | 0424 | Bright stars. |
| 52 | 0426 | Rich in the 10 cm. |
| 103 | 0432 | Fuzzy, 'arrowhead' shape. Not bad in 10 cm. |
| 69 | 0438 | |
| 70 | 0439 | |
| 54 | 0440 | Three globulars, seen clearly and identified in 10 cm. |
| 55 | 0446 | Large, but close to horizon. Seen with 10 cm. |
| 75 | 0458 | Faint in 10 cm. |
| 15 | 0459 | Bright. |
| 2 | 0515 | Moved 20 cm reflector to a site with better southern and eastern horizon. M2 is bright. |

| 72 | 0520 | Faint. |
| 73 | 0522 | A few stars. |
| 30 | | Just didn't rise in time. |

# 19   The sky on film

The sparkling clear night was one of the best I had ever seen. Since I had just spent several days adjusting my 20 cm telescope for photographic use, this would be the night for a first attempt.

There had been major changes to the telescope. I had mounted my first telescope, a 7 cm reflector, onto the side of the larger telescope, using specially made mounting rings. Then I had centered the main telescope on Spica, a bright star in Virgo, and aligned the small telescope so that it pointed in the same direction.

Instead of an eyepiece in the main telescope, I mounted a 35 mm camera body and focused it so that starlight would fall directly on the film, a procedure called *prime focus photography*. Centering the globular cluster Messier 2 in the smaller telescope, I opened the shutter and began a one-hour guided exposure. Keeping the cluster centered in the eyepiece seemed to be easy; it moved around a bit in the slight breeze, but generally the hour's work was pleasant. I took the film to the drug store, and several days later opened the package.

The film was absolutely blank.

Wondering whatever could have gone wrong was initially lost in a tremendous feeling of disappointment. I took the film home and examined it under a magnifying lens. The two or three spots I saw on the film could have been stars, or defects in the emulsion. The picture was an utter failure.

Or was it? Not quite; that blank film was a success! Thinking out the problem took some time, but I did learn from that disaster. The light breeze had a much greater effect than I had expected. The image was blowing all over the field, its light not getting a chance to record on a single spot on the film.

Also, I had no idea of the film's response to light. Had I begun by choosing a bright star to photograph, I would have known that the film was slow even for recording bright stars. Having used a slow Ektachrome with an ISO (International Standards Organization, the same as the old ASA) speed of 64, I should have known that the film was not fast enough to record faint astronomical objects. Combine that with the f/8 telescope, a little slow for recording of deep sky objects. I also should have known about this film's *reciprocity failure*, a characteristic that affects films to various degrees causing them to lose their sensitivity to light with long exposure times. No, I didn't leave the lens cap on, or forget to load the film, but I might as well have; all these other things probably produced the same result. That blank film was a fine teacher.

Every astrophotographer in the world has start-up difficulties with exposures that need to be guided at the telescope. There are so many things to remember – things like choosing the right film, removing the telescope's lens cover, finding the best object to photograph and the best time to observe, knowing how to guide the telescope accurately; it's a wonder any pictures come out at all!

Figure 19.1. The author's first astrophoto. Yashica A twin-lens reflex, 80 mm lens and Ektachrome X film, five-minute exposure.

The moral of this story is not that astrophotography is difficult, but that it should not be courted quickly and carelessly. If you want to capture the sky on film, it is best to start slowly and deliberately, work in steps, keep notes, and never give up.

## 19.1  Star trails

The world's largest telescope, it has been said, is simply a film of aluminum a fraction of a millimeter thick, and an equally thin layer of emulsion or electronics on which the image of an object is focused. The whole purpose of the massive pillars and blocks of metal is simply to keep these two thin layers the correct distance apart, and always pointed at the right place.

Thus, if it doesn't matter where the camera is pointed, astrophotography should be easy! And it is. For a start, try a star trail. The procedure is simple: find a dark sky site, put film with high ISO in your camera, lay your camera on a table or car roof with its lens pointed up towards the zenith, and open the shutter with a cable release for 30 minutes.

The results will be startling – a beautifully concentric series of curved lines! If you used color film, there would be the added beauty of different colors!

What you have is dramatic evidence of the Earth's rotation. While your camera was exposing film, the Earth was turning. Thus, the stars you have recorded are trailed. Your photograph has other evidence too – as we discussed in Chapter 1, the stars' different hues really do tell you something about their temperatures, with red stars being much cooler than blue ones.

The next step is a comparison. Try photographs of two different parts of the sky, one near the pole, the other near the equator in constellations like Orion or Scorpius. Notice how the 30 minute exposures reveal startlingly different lengths of trails; the trails in the polar one may be short enough that you can still identify the star patterns. Because stars near the Earth's axis of rotation have a much shorter apparent distance to travel in a single day of rotation, their trails are much shorter. The stars in the equatorial region are trailed much longer – and, if your camera is pointed near the equator, there will be two sets of concentric trails, one centered northward, the other southward.

If you use very fast film like Kodak's T-Max 3200, the bright stars will be overexposed if you leave the shutter open too long. With these films, try short exposures like 10 seconds to produce lots of stars with negligible trails.

A third stage of star trail photography involves actually choosing specific areas of the sky to photograph. If a half-hour exposure produces trails, a five minute exposure will also produce trails, but short enough that you can easily identify the stars and constellation patterns to which they belong. Now in short exposures, try shooting areas like the Big Dipper, the Orion region, or the southern constellation of Carina. Even though the stars will appear slightly trailed, the resulting image should be a good reproduction of what you can see in the sky. You will be able to identify constellations, smaller asterisms, and other details, and if you take enough pictures you may even develop your own star atlas.

Now try reshooting the photos that you have already taken, this time using different films. A fast color film like Ektachrome 400 will show more stars than Ektachrome 200. If you use a superfast film like Konica 3200 (which produces prints), you will find far more detail. Similarly, in black and white, trusty Verichrome Pan will record lots of star images, but Tri-X will record even more.

Next, try looking closely at the different films you have exposed, using a magnifying glass or a lupe. Compare the appearance of the images between the slower films and the faster ones. Does the faster film appear mottled, while the slower one is smoother and more natural? A consequence of increased film speed is an effect known as graininess, and you can see this grain as a mottled appearance in the faster films.

Meteors and artificial Earth satellites are a primary target for star trail astrophotography. With the fast films, there is a chance that your exposure will record a real prize, a bright meteor passing through! The chances for such an event are much greater on nights when meteor showers are present, especially in November when the Taurid shower produces bright fireballs (see Chapter 3). Remember that a meteor has to be brighter than 0 magnitude to record on most films.

If you photograph in the early evening or predawn hours, you may capture an artificial satellite. It is usually (though not always) easy to tell the difference between a satellite and a meteor in a photograph. Satellites usually leave long trails that carry through the picture; if they disappear into or appear from the Earth's shadow, their trails would show such a gradual change in light. Meteors tend to be more abrupt, their trails starting thinly, getting fatter, then disappearing. They also tend to be more irregular than those of satellites. Bright satellites are common after evening twilight and before morning twilight. Some show intensity variations which can be recorded on film.

## 19.2  The Sun

**Warning: Remember that viewing the Sun is dangerous! You must be careful of the Sun's light accidentally striking your eye.**

Much the same setup can be used for photographing the Sun as for viewing it. I have taken successful pictures merely by photographing the projected image; take an exposure or two faster than what your light meter suggests, and then take one or two slower. This procedure is known as bracketing your exposures; you choose the one that works best. If you use a filter that covers the telescope's front end, use a fairly fast film with an ISO of 200, keep your exposures around 1/125 second for sharper images, and bracket!

# 19.3 Moon and planets

It is possible to photograph bright solar system objects in the same way that I tried to photograph M2, with the camera at *prime focus*. Try an exposure of 1 divided by the film's ISO at f/16, or 1 divided by twice the film's ISO at f/11. If you are going to get seriously into photography you should experiment until you find out what is most satisfying to you. So, shoot lots of film and don't forget to record all exposure information such as time, telescope data, film, seeing, and transparency.

### 19.3.1 Photography by projection

Photographing bright objects like the Moon is not difficult in a process called *eyepiece projection* where the camera is not placed at the focal point. Instead, the eyepiece is at the focal point, just as for visual observing, and it projects the light of your object, like the Moon, into the camera body which is mounted a few inches away. In eyepiece projection the camera does not have its own lens. In a related method, called *afocal projection*, you would use the eyepiece as well as the camera's own lens set at infinity. Try a series of exposures, from 1/30 to 2 seconds, using moderately fast film.

Even though planetary photography is not guided, it is also not easy. For any real detail, use the highest magnification you can get away with for photographing the planets. Start with your ordinary camera, at prime focus, just as you did for a deep sky object. To increase the power, you may want to add a *Barlow lens* to the T-mount adapter. This lens increases the apparent focal length of the system, resulting in a larger image.

# 19.4 Guided astrophotography

Now that you have stepped lightly into the astrophotographic ocean, why not try going deeper? With long exposures of faint objects, we have to guide the camera, a greater challenge.

### 19.4.1 Camera support

Although mounting a camera on a tripod is an easy way to get started, it is decidedly not the best approach for serious astrophotography. The great disadvantage is that long exposures require that the camera follow the motion of the Earth somehow, so that as stars cross the sky the camera will follow them.

The easiest way to do this is to mount the camera on top of a telescope that already can follow the sky. Such an equatorial mount has two axes that are adjusted in such a way that a motor driving one will cause the telescope to follow a star (see Chapter 4).

Some companies market inexpensive mounts that attach cameras to telescope tubes. They either come with a $\frac{1}{4}$ inch diameter stove bolt, the same type of bolt that camera tripods use, or a place to insert one. If you cannot find such a device, why not attach the camera yourself? Simply drill a hole in the telescope tube, and run a bolt through. If you attach the camera directly to the curved telescope tube, it will not be stable enough. An additional wood or metal piece is needed that has two points of contact with the telescope's tube.

Now align the telescope with the camera. Using some bright object like the Moon or a bright planet or star, center the object in the telescope eyepiece, and then adjust the camera so that it too is centered.

### 19.4.2  What you need

As you get deeper into astrophotography, you will need some additions to your telescope. Normally these can be obtained one at a time, as you gain experience. First, the camera. The best automatic camera on the

Figure 19.2. As part of an all-sky photographic survey with his 12.5 cm Schmidt camera, Damien Lemay took this exposure of the northern Cygnus region. The 'North America Nebula' is visible near the top. Schmidt cameras are capable of capturing large pieces of sky – several degrees wide – in a single exposure.

market might work if that is all you have, but it is not the best camera for this purpose since the long exposures and cold temperatures can cause the battery to run out, causing the camera to die at mid-exposure. A quality camera with manual controls is what you want. Next, a high power eyepiece with an illuminated reticle will allow you to keep a star sharply positioned in one place. Use this eyepiece with the main telescope, as your camera is mounted piggyback on top. Later, you might want to take actual photographs through the main telescope; for that you will need to mount a separate guiding telescope. Guiding telescopes are mounted using easily obtained mounting rings. Use at least a 7 cm reflector or refractor with as long a focal length as possible; 100 cm is a good minimum. This is needed to get sufficient magnification for guiding; a rule of thumb is that your guiding magnification should be 2.5 times the focal length of your camera or camera–telescope combination. Thus, for a 20 cm f/7 telescope with a focal length of 140 cm, the power should be at least 330. Most guidescopes that are sold commercially are too small.

With a device called an off-axis guider, you can avoid the hassle of a guide telescope altogether. It is simply a narrow tube with a prism that fits into the eyepiece holder, directing a tiny amount of the telescope's light to a guiding eyepiece near the camera.

Your telescope needs not only a motor drive but also a drive corrector with a control to speed up or slow down the drive rate. It also helps to have electric controls in declination; but this is not totally necessary so long as what control you do have is smooth and not abrupt and jerky. Rarely will your drive move the telescope so accurately that speeding ups and slowing downs do not occur; small errors in the right ascension gear and other problems result in the need for correctors. Even professional observatories use them.

Finally, when you are ready to begin taking pictures through your main telescope, you will need a standard T-adapter and tube to attach the camera to the telescope. You do not need the camera's own lens at this time, for the telescope will act as the lens, the star image being focused directly onto the film, without using an eyepiece. Since the camera is attached to the telescope so that the film is at the prime focus, this form of astrophotography is known as *prime focus photography*.

### 19.4.3 Aligning the polar axis

As your astrophotographic demands increase, you will want to spend some time aligning the telescope's mount on the pole. The need for accurate pole alignment cannot be overemphasized; it will make guiding your photos much easier. With a properly aligned mount you rarely have to guide in declination. If the mount is not properly aligned, the resulting effect will be that the stars at the edges of your photograph will show curved trails, an effect called field rotation. The more accurately your telescope is aligned on the pole, the better your astrophotos will be.

Although the procedures are somewhat cumbersome, you should accomplish them without too much difficulty. It will, however, take a little time.

One method involves moving the telescope in declination so that three stars of identical right ascension successively come through the field: Alpha Draconis, Delta Draconis, and Polaris are examples. Any three stars with the same right ascension can be used, and this method is effective in both hemispheres. If Alpha Draconis is too low in the sky, 50 Cassiopeiae should be well up. At first, the line will not be accurate, and the three stars will not appear in the eyepiece field during a single sweep in declination. But after gradual adjustments of the altitude of the polar axis, and moving the entire mount in azimuth, the mount should come into alignment.

Following this trial-and-error procedure may take some time. Another approach is called the drift method. Begin with a star near the celestial equator, and near the meridian. Using your highest magnification, turn on the motor drive and watch the star drift through the eyepiece. Never mind the east–west drift; see if the star drifts north or south. If the star moves to the north, the polar axis is set too far west of the pole; if it moves to the south, the axis is too far east. Do not adjust the altitude axis now; just move the scope in azimuth. Unless the telescope has a special azimuth adjustment, you will have to handle the entire mount to do this. Repeat the procedure until there is no north–south drift.

Now find a star on the equator but near the eastern horizon. Again see how this new star drifts. If it drifts southward, you have to increase the altitude of the polar axis to reach the pole. Decrease the altitude if the star drifts northward.

For portable telescopes, I recommend a polar axis finder such as the one sold by Roger Tuthill and advertised in the major astronomical magazines. These devices allow a precise offset by 0.9 degree, the distance between Polaris and the north celestial pole, and with an accompanying table you can quickly align the pole using Polaris. Even with this device, you may still want to make final adjustments using one of the classical methods.

If your motor drive is working and if the telescope is reasonably well aligned on the pole, a 10 minute exposure through the camera's ordinary lens will produce sharp images of stars. Your homemade atlas of constellations will now consist of well guided stars, not trails.

If you want your pictures to be even better tracked, you might want to help the motor drive along by guiding the telescope yourself. You need not try this unless your telescope has some sort of slow motion controls. Choose a pattern of stars such that several stars are around the edges of the field of view. After a minute you may notice one of the edge stars move out of the field. Using the slow motion control, move the star pattern back to where it was when you started the exposure. If you have aligned your telescope on the pole carefully, you will hardly have to correct for declination shift at all. (At low elevations, atmospheric refraction will cause declination drift for even a perfectly aligned mount.) Later you may want to buy a special guiding eyepiece with illuminated crosshairs; this will allow you to guide accurately on one star in the middle of the field.

### 19.4.4 Setting up the picture

With your camera mounted on a well-aligned telescope, you are now almost ready for any sort of guided astrophotography. Set your telescope on the object you wish to photograph. Next, do *not* open the camera! Instead, try guiding for a minute just to make sure that your system is tracking properly. Is one of the illuminated crosshairs aligned east–west? If not, rotate the eyepiece. If the star stays in the center of the field, and you know which button to press to correct the position a bit, then you are almost ready to start your exposure.

### 10.4.5 Focusing

For bright planets or the Moon, focusing the camera is easy. If your target is a faint nebula or cluster, you may have trouble seeing it in the camera's viewfinder. To focus, it will be necessary first to locate a bright star and focus on that star, then move to the faint object to be photographed.

You also need to choose an appropriate guide star, preferably not too bright that it will swamp the guiding eyepiece, and not so faint that you cannot see it. Second or third magnitude stars are usually very good for small guiding telescopes.

### 19.4.6 Ready!

After all this preparation, you probably want to hold your breath as you finally open the shutter for your exposure. Even now, there are several things to remember. If it is cold, the film may tend to get brittle, and might even break inside your camera. Also, you might be wearing a heavy-duty nylon observing suit that rubs around you as you move, building up a charge of static electricity that is released as you touch the camera, with lightning on the film as a possible result. If the air is damp when your night is over, try placing the camera into a plastic bag before bringing it inside. This way moisture will condense onto the outside of the plastic bag, not on the delicate camera and film.

Now open the shutter but have the telescope's aperture blocked by a piece of black cardboard. Allow the vibration of opening the shutter to die out. Finally, remove the cardboard and start the exposure. You can use the same 'hat trick' to end the exposure.

## 19.5  Some advanced ideas

As you gain confidence with your approach to astrophotography, you may try some special procedures and devices. The following ideas do add extra work to an already busy routine, but they do produce interesting results.

### 19.5.1  Copying

By simply copying your color slide or negative onto a slow film like Ektachrome 100 or Kodachrome 25, you considerably increase the contrast, often bringing out details that were not obvious in the original. When you attempt this, remember that the copy will obviously not add anything that is not in the original; it acts as a process to enhance what already appears.

### 19.5.2  Hypersensitizing

With heat, and a 'forming gas' combination, usually of 92% nitrogen and 8% hydrogen, films can be rendered so much more sensitive to light that a whole new vista is opened to you. At some point you really should try this, but as with the rest of astrophotography, proceed in steps, At first you would be wise just to order 'hypered' film from a company (like Lumicon) that offers it, then store it in the freezer. If you then plan to get into hypering yourself, you can buy or build a hypering tank into which film will be placed. The tank is pressurized and flushed, pressurized again, and then the film is baked for several hours. You then put the film in the camera and proceed as usual.

## 19.6  Processing film

The easiest way to develop film, of course, is to take it to the drug store or chemist's. With some caveats, that approach is fine. Expose the first and last frames of the film on some daylight object, so that the laboratory knows where to cut the film. To add insurance that M42 doesn't come divided, also ask that the film is not cut in any place.

Developing film yourself is fun, but if you are not used to doing that, why not get proficient at one thing at a time. Get good at taking pictures at various exposures, then worry about processing.

A darkroom is not as frightening to set up as you might think. It is simply a light-tight room; black poster paper should be able to block window light in a bathroom, or use a large closet.

The procedures for developing films in trays or in developing tanks are well described in books on introductory photography, and the times recommended for processing are also printed on film instructions. You will usually use what is known as the *time and temperature* method in which three chemicals, the developer, the short stop, and the fixer, are raised or lowered to a specific temperature, and then used on the film for a certain length of time. When the negatives come out of the last chemical, you rinse them and then examine them with a magnifier, and evaluate your results. Is the image focused? Is it well guided? Is the sky background dark? If this set of negatives

Figure 19.3. In this difficult shot, Terence Dickinson captured a bright waning crescent Moon with a faint and slightly fuzzy Comet Okazaki–Levy–Rudenko to the top. 15 cm f/6.5 refractor, 30 seconds on Agfachrome 1000.

Figure 19.4. The Moon with some stars from the Pleiades, photographed by Steve Edberg on June 12, 1988, 10 s Ektachrome 400, with a 140 mm f/3.64 Schmidt–Newtonian telescope.

isn't satisfying, the next should be better; the prime rule of astrophotography being to keep on experimenting and never give up. Always take notes at every step so you can learn from your own experience.

With all this information, you are likely to be certain that as soon as you take your first astrophotograph everything will turn out perfectly. Your picture of M42 will turn out to be the finest picture ever taken of anything.

Well, not quite. Even with all that preparation, you still need to experiment. Try different films. Try different exposures. Try different locations. You will probably spend at least a year experimenting before you are finally satisfied with your pictures, but you will feel the expertise coming and you will have a lot of fun.

## 19.7  Some hints

1. Make sure your mount is sturdy enough. Recently I started observing with a Schmidt camera, a very fast system capable of photographing wide fields of view with considerable speed.

I found that my moderately steady mount, which was fine for visual observing, was absolutely useless for photography of deep sky objects. If I so much as touched the setup, or if a light gust of wind came along, my exposure was ruined and the stars looked like jagged lines!

2. The motor drive needs to be of good quality. Visually, it does not matter if M13 floated from side to side in the field; but the camera, as we all know, is far less forgiving, and in my early pictures all the stars are trailed!

3. Many high-tech commercial mounts have little red lights that conveniently turn on to let the observer know that the drive is powered. That is very nice for visual work, but to the camera the 'little' red light is blinding after a 15 minute exposure!

4. Watch for stray car lights, street lights, and moonlight. Fast film fogs!

5. Be patient. Celestial objects are generally faint, so that long exposure photographs are needed. This process can be very difficult. There are so many things that could go wrong that you really should not expect success until you have tried for a while. I have seen people with no experience come into a telescope store, look longingly at a photograph of the Andromeda Galaxy taken with the Mount Palomar Telescope, and decide to 'buy a telescope so I can take a picture like that one'.

6. Get to know your telescope. Practice first, using the right ascension (RA) controls and those of declination (Dec) until you are truly at home with guiding the instrument. If your scope is very slightly off in Dec, you will still be all right as long as you know the direction in which to compensate.

7. Beginning astrophotographers should stick to a single basic film. Get to know it well; find out its advantages and limitations. Kodak has introduced a film, technical pan 2415, that offers high contrast and has given excellent results for solar, lunar and planetary photography. For deep sky, Ektachrome 400 yields decent results, especially if slightly green sky back-

grounds don't bother you. If possible, process the film yourself. Otherwise you could get your favorite constellation cut down the middle.

Guided long-exposure astrophotography is not for the faint of heart. It does require a commitment of patience, a satisfied knowledge that you are in it for the long haul, and that you are not expecting your first results to be Rembrandts.

# PART FIVE

# Special events

## 20 Solar eclipses

It was a special moment by a snow-covered country road when the Moon's first tiny bite out of the Sun signalled, right on time, the beginning of the eclipse of February, 1979.

The increasing darkness of the partial outer shadow or *penumbra* caught up with me gradually. I started thinking that a light cloud must be covering the Sun, but no; through the protective glass I could see half a Sun clearly. By the time 80% of the Sun was covered, however, the darkness was increasing rapidly and our shadows, though dim, were razor sharp. The breeze stopped, it got much colder, and the landscape seemed to develop a sense of foreboding. It would be understating my feeling to suggest that I was thrilled to be a fourth body in a cosmic lineup of Earth, Moon, and Sun. Never before have I felt that I was an actual part of a cosmic event, but the emotions that were beginning to surge inside would not deny it. And still the sky got darker; there would be no escaping what was to happen. Towards the south I could see the black umbra rushing at us, and before I had had a cnance to look again, the Sun had disappeared.

In its place was a jewelled crown. The Sun's outer atmosphere, its *corona*, was bright and almost circular. There were several *prominences* or tongues of material jutting from the Sun's edge; one was quite large. I finished the photography within 90 seconds as planned, but I was not really conscious of my work. I was transfixed by the spectacle. It wasn't just the *Sun*, either. The *sky* was very dark. The *Earth* was still as a painting. An hour before noon, nearby street lights were on. And enveloping the whole scene was an unearthly hush.

During your lifetime you must try to see a total eclipse of the Sun. I have yet to see a spectacle that rips to the core of my being more effectively than such an event. (Discovering a comet also has an extremely powerful effect, but much different from that felt when seeing a total eclipse.)

239

So far I have seen only three *central* eclipses, during which the Moon passes directly in front of the Sun. Two of these, on July 20, 1963, and February 26, 1979, were *total*, where the Sun was completely covered. The third, on May 30, 1984, was *annular*, where the Moon is so far from the Earth that it does not completely cover the Sun, leaving instead an annulus or ring. In a very limited sense it doesn't matter if I never see another, for the memory of those events will never leave.

## 20.1   Alignments

Three types of special events involve two bodies aligning with each other as seen from the Earth. An *eclipse* occurs when these two bodies are roughly the same apparent size, as the Sun and Moon are. An *occultation* occurs when a large body, like the Moon, passes in front of a smaller body like a star or planet. Chapter 13 on asteroids considers an asteroid passing in front of a star. It seems preposterous to think of a five kilometer wide piece of rock as being larger than a million-plus kilometer wide star, but the asteroid is close enough that it has finite angular size, compared to the star.

Figure 20.1. Total eclipse of the Sun, February 16, 1980. Steve Edberg used a 485 mm f/5.5 lens and Ektachrome 400 film at 1/15 s for this photograph. Notice the circular shape of the corona, which is typical of times around sunspot maximum. When the sunspot cycle is at minimum the corona is of a more oval shape.

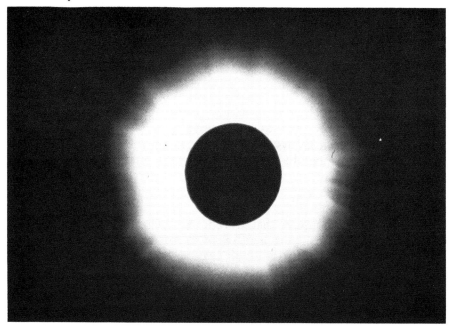

Although the mechanics of such alignments, involving two bodies in orbit, are easy to understand, the effect of the resulting event is enormous. This is not surprising; anyone who has seen a total eclipse of the Sun will understand why eclipses have terrified generations of people. Even today there are few things as thrilling to the general public.

Other chance alignments have become fodder for astrologers who have seen them as having a predetermined effect on our lives. There are planetary appulses, where a planet approaches a star, and a conjunction involving two planets simply appearing to be close to one another. While these events have no value in the sense of collecting scientific data, they are fun to watch.

Chapter 12 discusses events called *transits*, where a small body (a planet) passes in front of a larger body (the Sun).

## 20.2 Solar eclipses and the public

The school clock turned to 11 a.m., the bell rang loudly and the principal walked from room to room checking to make sure that all blinds were tightly closed and that the minds of the children were drawn away from the tightly covered windows. Some teachers were showing movies; others were leading their classes in song.

A nuclear holocaust? Civil defense drill? Bomb scare? No. Then what was so terrifying that all children had to be protected from it? This was an eclipse of the Sun: in February, 1979, scenes like this occurred all over Canada. And in Winnipeg, where the eclipse was total, a school superintendent justified this action on the curious grounds that 'eclipses have no educational value'.

Such a response to a spectacular show given by our star and our satellite reflects the barbaric terror in the minds of earlier civilizations. Of course the Sun is dangerous, but NOT just during a total eclipse. It is always a bad idea to look at the Sun with unprotected eyes, even for a moment. And an unprotected look of just one tenth of a second will cause permanent eye damage. But the answer is not to ignore the habits and geometrical alignments that involve the always changing Sun, rather it lies in education and in learning how to see and to study the Sun with safety.

### 20.2.1 Eye protection

The blindness scare that precedes any eclipse of the Sun, partial or total, often creates an atmosphere of uninformed hysteria. Amateur astronomers are in a good position to overcome this. Public astronomical demonstrations and education, both directly and through the media, before an eclipse can be of great help in making people aware of both the real dangers of the Sun and effective ways to overcome the danger.

A pre-eclipse presentation to school groups, teachers, parent–teacher associations, district superintendents – anyone involved in education – should emphasize the following points:

(1) The Sun is always dangerous. No special rays come out during eclipses. It is only that the partially eclipsed Sun attracts more attention.

(2) You do not have to lock children up during an eclipse. With demonstrations about the power of the Sun (see Chapter 8 for a description about burning a cheap garbage bag) and a lot of supervision, large groups of children should be able to watch an eclipse with safety. Parents should join the teachers in the supervision to make it more effective.

(3) There are safe ways to watch an eclipse:

Projection through a telescope or binoculars.

Projection through a pinhole, but do emphasize not to look through the pinhole!

Use welders glass no. 14 or higher.

If you are in the path of totality, when all the photosphere, or the part of the Sun we normally can see, is covered, no protection is needed. But remember to resume protection the *instant* the first rays of the Sun peek out from behind the lunar edge.

My parents argued this point with me during the 1963 eclipse. At the time I was living in Denver, and was able to travel 2000 miles home just to see the eclipse. Unfortunately the articles we read failed to convince them that totality would block all dangerous rays. I promised to be careful, and as the Sun began to turn into a thin sliver, we marvelled that we could see the progress of the eclipse by looking at the Sun's projection through spaces between leaves – hundreds of crescent Suns were dancing in the shadows. With the onset of totality, the glass I was using blocked all light, and I could see nothing at all! Without thinking I ripped off the protective covering and looked at the eclipse. My parents and I saw a most stunning event that afternoon, but they were concerned about the lack of protection.

They need not have been. Astronomers have looked at the total phases of solar eclipses with telescopes, and without protection, for centuries, without ill effect. The tenuous rays of the corona are much too weak to cause the slightest harm. What is absolutely essential, however, is that you do not remove your protection until second contact, the point at which the Moon completely covers the Sun, occurs. It is pretty easy to tell when that happens; as long as you can see any sunlight through the welder's glass, it is not safe to remove it. If you are observing the partial phases by projecting the Sun onto a piece of paper, you may look at the projected image all you want, of course, but do not look directly at the Sun. As the crescent Sun gets smaller the darkness rushes towards you as a wave. You can virtually feel the darkness, and then you will suddenly no longer be able to see anything through your glass. At that point, it is safe to look.

Remember: the entire total phase of the eclipse, when absolutely no bright photosphere is visible, is safe to watch. At the onset of totality, the corona near the edge of the Sun that has been most recently visible is brighter. As totality deepens, the corona near the edges becomes more equal in brightness. Then you will see a brightening on the opposite edge, and suddenly a flash of light, the diamond ring. The instant you see the direct light of the Sun's photosphere, it is no longer safe – protection is needed again at that instant.

## 20.3   The saros cycle

When the Moon's path through the sky intersects the ecliptic, that point is called a node. If the Moon is moving northward the point is called the ascending node; if southward, the descending node. If the Moon should be near one of these nodes at its full phase, a lunar eclipse can happen. If this happens at new phase, then a solar eclipse would take place.

The Moon orbits the Earth in a peculiar pattern that rotates in space over a period of time as the Earth–Moon system orbits the Sun. For an eclipse to happen again, the Moon must be in the same position in its orbit relative to the same node, the Sun and Earth must be the same distances from Earth, and these conditions must be at the same time of year. This alignment repeats every 18 years, $10\frac{1}{3}$ or $11\frac{1}{3}$ days. This period is known as a 'saros cycle' and is the reason eclipse predicting has been successfully done for so many centuries. The extra third of a day means that an eclipse would happen again at a different longitude.

The December, 1974, issue of *Sky and Telescope* included the complete saros information for the December 13 partial eclipse, the 12th in its saros series. The first eclipse of that saros was a very small event visible only from the Arctic on August 14, 1776. There was a small partial eclipse on December 2, 1956 (the 11th in the series), and the 13th, on December 23–24, 1992, will be a stronger partial eclipse that crosses the international date line.

On February 28, 2101, the first central eclipse of this series will occur, but it will be only an annular. It will not be until May 16, 2227, that the first total eclipse will take place. About half a millenium later, the series will close with a small partial eclipse in the Antarctic. Many saros cycles run concurrently, yielding at least two solar eclipses, visible somewhere on Earth, each year.

## 20.4   Partial eclipses

If the Moon passes completely in front of the Sun, the resulting eclipse is total or annular. If it does not pass completely, then the eclipse is *partial*. The event consists of the Moon moving across the Sun between the first and last contacts of the Moon's edge with the edge of the Sun.

Partial eclipses of the Sun are not very 'busy'. A partial eclipse begins when the Moon takes its first bite out of the Sun, and, as minutes go by, more and more of the Sun gets covered. After the maximum point of coverage is reached, the Moon continues on its way, releasing the Sun to its former glory.

You can, of course, time accurately the first and last moments of lunar–solar contact (although you need to be really alert to catch first contact), as well as the Moon's passage over sunspot groups. But other than that, partial eclipses are not very useful for visual observers. They are mainly for you to enjoy, either with friends or with school or public groups. Actually, a partial eclipse provides a fabulous opportunity to organize a community party in a local park, or even in front of a civic building. For the partial eclipse of December 13, 1974, a group of us set up a telescope at the front entrance of McGill University's Arts Building, the single most crowded spot on that whole campus. People would be able to get a between-class look at the eclipsed Sun projected on a piece of cardboard. Surprisingly, even though the sky was completely cloudy from first contact to last, the event was a success! We introduced hundreds of people to the art of looking through a telescope, and publicized our astronomy club.

## 20.5  Total eclipses

The number of things to do during a total eclipse increases rapidly as totality nears:

**Natural pinholes** If your observing site is near a tree, you will see an interesting event occur as pinhole images of crescents are projected by spaces between leaves on trees. Try photographing this effect. To ensure that one of your pictures turns out, take a number of pictures at different exposure settings (bracketing, a procedure discussed in Chapter 19).

**The darkening sky** The closer the onrushing lunar shadow gets, the darker your sky will become. As totality nears, the sky will darken as fast as if someone were lowering the light level with a dimmer switch. This effect also is interesting to photograph, this time with the camera's normal or wide-angle lens. Begin with the correct exposure for the normal, sunlit scene, and do not change the exposure as the sky begins to darken. Closer to totality, the sky will darken so much that you will have to adjust your exposure to get anything at all on your film. You will be left with a wonderful photographic memory of a darkening sky.

**Shadow bands** A minute or two before and after totality, lines of darkness sometimes pass by as bands on flat surfaces. This phenomenon is much easier to see if you are watching the eclipse from a field of snow. Placing a white cardboard screen on the ground might help make these subtle bands more visible.

**Baily's beads** As the last rays of disappearing sunlight shift across valleys on the Moon's edge, the effect of beads is very striking. First described by Francis Baily, a stockbroker and amateur astronomer who observed the phenomenon during an eclipse in 1836, the beads last just a few seconds right before and right after totality. If the Moon's apparent diameter is just slightly smaller than that of the Sun, the resulting annular eclipse might produce Baily's beads all around the almost-covered Sun, an effect called a *diamond necklace*.

**Diamond ring** For no more than a few seconds, a bright spark of photo-sphere combines with the corona to produce a 'diamond ring'. The effect is usually more memorable after totality. Our eyes are unfiltered, and the effect of the gradual brightening and the sharp flash of light is startling. But a second is all you get; filter your eyes as soon as you have seen the diamond ring that signals the end of totality.

### 20.5.1  Photographing a solar eclipse

The idea of going to a solar eclipse without a camera seems absurd, but the question of how to photograph the event is certainly not. With so

Figure 20.2. Baily's beads during the February 16, 1980, eclipse of the Sun. The curved bright line on either side of the brightest bead is part of the Sun's chromosphere. 1250 mm f/10 Schmidt–Cassegrain telescope, Ektachrome 200 film at 1/500 s. Photo by Steve Edberg.

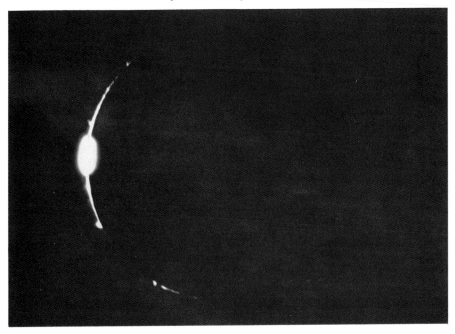

much happening in so short a time, it seems a natural idea to want to record as much of it as possible on film or videotape.

It actually is easier to consider videotape first, for what you see in the camera's viewfinder monitor is what gets recorded onto the tape. Unfortunately, the camera's normal lens will not record much of the Sun; however, a telephoto lens should record the eclipse superbly. A number of films, like Kodak's Ektachrome 200 or 400, are effective for eclipses. To get the partial phases exposed correctly, so that sunspots are visible, try some test exposures a few weeks before the eclipse. As long as any of the Sun's brilliant photosphere is visible, use a strong (no. 10 or higher) Welder's glass filter to cover your camera. As totality begins, take the filter off and record what you see. If you have taken many different lengths of exposures, you can expect each to show a different amount of the corona's soft light.

With film, almost anything goes during totality. The shot that overexposes the inner corona and prominences might record the fainter outer parts beautifully. It is important to have a fresh roll of film ready just before totality begins and take lots of pictures at increasing f/ratios or increasing shutter speeds. One idea is to take lots of pictures of the pinhole effects or landscape about 15 minutes before totality, then change film so that you have a full roll for the total phase. Make sure you have enough film, and don't get caught changing film two minutes before totality!

**The prominences** The pink tongues of flame that appear at the Sun's limb during a total eclipse are actually all over the Sun's surface; we see them as prominences on the edge, or filaments on the disk, with a hydrogen alpha filter. It is only the ones at the edge that we get to see without any filter at all during a total eclipse. Although it is possible that the appearance of a prominence can change during the short period of totality, the changes you are likely to observe are caused by the Moon covering differing amounts of a prominence's base. Since these are brighter than the corona, the shorter exposures you take are likely to record them best. There would be a greater number of prominences at sunspot maximum.

**The chromosphere** Other than the prominences, the pink, thin, inner atmosphere of the Sun is the brightest part of the visible complex during totality. If the Moon is not excessively larger than the Sun, it might appear as the background to the prominence shots you take.

**The corona** Although the broad outer atmosphere of the Sun reaches a temperature of millions of degrees, it is so tenuous that a spacecraft passing through would hardly feel it. The corona is the real showpiece of a solar eclipse. During sunspot minimum, its shape is usually sharply elongated, and near maximum it tends to be more rounded. Expect to see the corona extending to a full solar diameter around the eclipsed Sun!

**The environment** Remember that old camera you haven't used in years? Why not bring it along in addition to your new one? Then, while you are taking telephoto shots of the corona, your old camera with its regular lens can be shooting the eclipsed Sun surrounded by a dark sky – a most striking effect.

   Practice first. Set up your cameras ahead of time and run through the procedure you plan to follow. Take pictures at various exposure settings, using your second camera to record the sky near the Sun and other environmental effects.

## 20.6   Other activities

   **Bright stars and planets** If the eclipse is long enough and you can divert your attention away from the corona, try looking for nearby planets, especially Mercury or Venus. Often magazines like *Astronomy* and *Sky and Telescope* will publish a map of the sky around the eclipsed Sun that will include nearby planets as well as stars.

**Variable stars** Variable star fanatics often use eclipses to record estimates of bright stars that have been lost in twilight and the solar glare for weeks. Otherwise these stars would not be seen until weeks later. To miss the beautiful eclipse just to observe a distant variable star is something I would hardly recommend, however.

**Comets** Sweeping the sky for comets is an activity: on rare occasions bright comets happen to be closing in on the Sun just at the moment of totality. A late nineteenth century comet known as 'Tewfik' was within hours of its closest approach to the Sun when it was found during a total eclipse. It was never seen again. In 1948 a bright comet, known as the Eclipse Comet, was found in much the same manner, although it was seen in the night sky afterwards. The flurry of discoveries of sungrazing comets by the Solwind and Solar Maximum Mission satellites indicates that comets of this type may be more common than previously thought.

**Edge of path** Some observers place themselves directly on the edge of totality so that the width of the Moon's shadow can be measured accurately. The advantage of going two or three miles inside the path may be an unparalleled view of Baily's beads and an extended view of the chromosphere. If total eclipses took place every month, this activity would be worth serious consideration, since important data can be obtained this way. However, if you are considering doing these things, you need to consider whether they are worth the possible price of missing the entire total phase of the eclipse.

**Spectra** If you are interested in capturing spectra on film you may want to try photographing the Sun's 'flash spectrum' using a camera and spectroscope or a simple prism at the moment totality begins or ends. Normally all the spectroscope will show us is the 'absorption spectrum' of the Sun's bright photosphere. During totality we see an 'emission spectrum' of the chromosphere, or inner atmosphere, and then the mixed spectrum of the inner corona. At the instant of second contact, when the Moon first fully covers the Sun, and again at third contact, the spectrum changes from one type to the other, and this sudden effect is called the 'flash spectrum'. With a small telescope and camera, amateur photographer Steve Edberg has successfully captured the flash spectrum during the February, 1980, and March, 1988, eclipses.

## 20.7  Annular eclipses

Annular eclipses occur when the Moon is too far from Earth to cover the Sun completely, so that during the central phase the blackened Moon is surrounded by a ring of sunlight. On May 30, 1984, the Moon and Sun were so close to being identical in size that at some places along the path observers could see a ring, while at others all that was visible was a 'diamond necklace'. I observed this eclipse from a site near New Orleans, and was surprised at how many of the effects of a total eclipse were present. The wave of darkness was especially moving as the Moon shadow rushed towards us from the southwest. In fact, it got as dark as it would have done for a total eclipse, with an eerie twilight sky, but the shadow covered us for a very short time before racing on. In a real annular eclipse, the ring of sunlight would prevent the sky from darkening very much.

## 20.8  Enjoy it!

The two or three minutes that you might have to see a total eclipse are very rare. On Mars you could see a transit of the Sun by Phobos, one of its moons, but the Sun would be so much larger than the Moon that you might scarcely know the event was happening. Or a different combination of moon and planet would produce an occultation so long that the Sun and its corona would disappear for hours. Only on Earth is the combination so perfect that the Moon can cover the Sun just enough to bring out the magnificence of the prominences and the corona. So delicate is this event that the longest it can last is about seven minutes. This rare and special period of totality offers a moment to think the unusual thought that only at a particular place and time can a four-body arrangement of Sun, Earth, Moon, and observer produce this special event. It is a chance to show how you belong to the universe.

Whatever you plan to do for a total eclipse, don't let photographing the event, or anything else, obscure your chance of just standing up and looking at it. During the 1963 eclipse I drew the shape of the corona. My fondest memory is of just looking up and watching.

I do not plan to have a camera to record the eclipse on July 11, 1991. I plan just to watch the cosmic alignment. For me, it is important during those minutes to *feel* that I am experiencing something precious.

# 21 Lunar eclipses and occultations

At a Lunar Eclipse

Thy shadow, Earth, from Pole to Central Sea,
Now steals along upon the Moon's meek shine
In even monochrome and curving line
Of imperturbable serenity.

How shall I like such sun-cast symmetry
With the torn troubled form I know as thine,
That profile, placid as a brow divine,
With continents of moil and misery?

And can immense mortality but throw
So small a shade, and Heaven's high human scheme
Be hemmed within the coasts yon arc implies?

Is such a stellar gauge of earthly show,
Nation at war with nation, brains that teem,
Heroes, and women fairer than the skies?

Thomas Hardy, 1903

## 21.1 Lunar eclipses

Even though it would initially appear that lunar eclipses are far less inviting than their solar equivalents, I can look back on many beautiful events, the eclipse of December 30, 1963, particularly so. A sharp cold front had rushed through, leaving the Montreal sky clear and bitterly cold. The eclipse began with a barely detectable penumbral shadow; we could not detect any darkening at all until this shadow was almost half way across the face of the Moon. During the last few minutes before the arrival of the Earth's umbra, or dark inner shadow, the Moon was significantly darker.

The umbra attacked the Moon like an onrushing army. Our satellite was literally disappearing; where was the dull red glow we had expected? The

Moon had shrunk to a thin crescent which narrowed and finally disappeared altogether.

The eclipse was now total, and later we would learn it was one of the darkest on record. Through a telescope we were able to see the Moon's dim outline, but we needed a star chart to find it.

The following December a group of high school friends observed another total lunar eclipse. This time the weather was warmer, and we decided to use a tape recorder to record our observations. As the shadow crossed the centers of the small craters, we noted the time on a tape recorder; for large maria we noted the times of its passage past each wall. Unfortunately we did not have the recorder on long enough for the shadow crossing at an edge of the Sea of Crises, and missed recording the time. When we played it back around December 25, all we heard was a cheery 'Mare Crisium!'

A truly patriotic observing team, we ended our session by singing the national anthem.

### 21.1.1  Shadows

When the Earth moves between the Sun and the Moon, an eclipse of the Moon happens. Unlike solar eclipses, which are concentrated on a small strip of the Earth's surface, a lunar eclipse will be visible throughout the entire hemisphere where the Moon is in the sky. Thus, although lunar eclipses do not occur as often as do solar eclipses, they are more widely observed. Eclipses may be total, in which the Moon completely immerses itself in the Earth's shadow; partial, in which only a portion is covered; and penumbral, where only the Earth's outer shadow covers a part of the Moon. Unless the Earth's main shadow, or umbra, is close to touching the Moon, the penumbral events are hard to detect.

Each of us also has umbral and penumbral shadows. Look at your shadow on a bright day. The part that completely blocks off the Sun's light is your umbra. But notice that your shadow's edge is somewhat diffuse, fading out gradually from total covering of sunlight to no cover at all. That narrow area of diffuseness is your penumbra.

Little explanations like this are fun to give to the public, and since lunar eclipses are special events, a wise amateur astronomy club might take advantage of their popularity to organize at least two observing programs. The first could be a public 'eclipse party' in front of your local planetarium, library, school, shopping center, or at a safe and popular park. Just as solar eclipses can inspire the public into an interest in astronomy, a lunar eclipse can too, and without the accompanying safety concerns. Observers and telescopes show the Moon to passersby, just as you may have seen it done some time ago. This is a fine way to publicize both your hobby and the group that supports it.

The second project involves detailed observation of the eclipse, and if your club has enough people, it is a good idea to make the two observing sites separate. You would be surprised how many people would rather enjoy an eclipse at a public party, leaving other observers to make detailed observations.

### 21.1.2 Things to do

**Color change** As the Earth's shadow cuts deeper, the full Moon will naturally darken. However, it may also show evidence of color change, perhaps ranging from a bright red to a duller brown. These changes are often seen through binoculars or even with the unaided eye. Record details of any color changes that you might notice, including the telescope and eyepiece you used for the observation.

**Craters** Occasionally some of the larger craters, or walled plains, might look brighter than the surrounding terrain while in the Earth's shadow because the material on their floors reflects light more effectively than the surrounding terrain. Note changes in color intensity.

**Darkness of totally eclipsed Moon** One of the best ways to judge the dust content of Earth's atmosphere is to watch the Earth shadow's behavior during a lunar eclipse. Some eclipses are so bright that an orange shadow is surrounded by an almost white rim. On the other hand, the December, 1963, and December, 1982, eclipses were so dark that the Moon was almost invisible. The cause of this variation is largely the atmosphere's dust, the amount of which depends partly on volcanic activity. A single major eruption the size of Mt St Helens in 1980 can darken an eclipse that occurs months later. The scale devised by A. Danjon is normally used for determining this level of darkness. Its range is described thus:

> L=0. Very dark eclipse, Moon almost invisible, especially at mid-eclipse.
> L=1. Dark eclipse, gray or brownish color, details distinguished with difficulty.
> L=2. Deep red or rust colored eclipse, with a very dark center of shadow. Outer edge of umbra is relatively bright.
> L=3. Brick-red eclipse, usually with a bright rim, sometimes yellowish, to the shadow.
> L=4. Very bright coppery-red or orange eclipse, with a bluish very bright shadow rim.

**Magnitude of eclipsed Moon** In the same way that some amateurs estimate variable stars, it might be possible to estimate the magnitude of the totally eclipsed Moon by comparing its brightness with those of bright stars. If possible, choose bright stars close in altitude above the horizon to the Moon, so that different amounts of atmosphere will not interfere with your estimate. You will probably find that its magnitude is somewhere between −4 and +3, as opposed to its usual uneclipsed brightness of −12.5. Since our sky is not crowded with −4 magnitude stars, magnitude determination will be difficult unless the eclipse is a dark one; during the December, 1963, eclipse someone estimated the Moon at +5.

**Stars visible during totality** Compare the numbers of stars visible with the unaided eye before the eclipse begins, during the early and late partial phases, at the onset and end of totality, and at mid-eclipse.

**Shadow contacts** A favorite activity is watching the Earth's shadow march across the Moon's face, covering familiar features as it goes. If you have already observed or drawn some of these craters (see Chapters 6 and 7), seeing them in this way renews old friendships. Keep a record of the times the shadow crosses these various features, and try some quick sketches to show the shadow as it crosses craters.

It is possible to time shadow contacts on craters and the limb of the Moon. Because the Earth's umbral shadow is not as well defined as that of the Moon, these shadow contacts are difficult to time more accurately than to the nearest half minute.

Try to record which major features – maria and craters – are visible, and are not visible, during totality.

**Photograph the entire eclipse** For the eclipse of May, 1975, I tried photographing the entire event on two or three strips of film. For this project I used a camera that would not automatically advance the film after each shot. The camera was mounted on a tripod with the Moon in mid-penumbral phase, and positioned on the east side of the first exposure. Taking exposures every five minutes, I wound up with a very nice report of the entire eclipse on only three frames. The results are shown in Figure 21.1.

**Detailed lunar eclipse photography** As for eclipses of the Sun, taking pictures of a lunar eclipse is a good idea. You should use color film, for the shadow will usually display enough color to record. Exposure settings depend on whether you use prime focus or eyepiece projection photography (described in Chapter 19), and on how dark the shadow happens to be; thus an exposure guide will not be very helpful. With ISO 200 film, take a wide range of shots from about 60 seconds to 1/125 second depending on the eclipse phase and darkness at totality. Film is cheap; use lots!

Figure 21.1. The lunar eclipse of May, 1975, photographed by the author.

**Time and weather record** Keep a record of all these events, using short wave time signals. If a short wave signal is not available, at least compare your watch to an accurate signal before observations begin and after they are over.

### 21.1.3  Penumbral eclipses

A significant fraction of lunar eclipses are only penumbral, in which the main shadow of the Earth never touches the Moon. If almost all the penumbra crosses the Earth's shadow, you are likely to see an odd shading effect on the part of the Moon nearest the umbra; if less than half the penumbral shadow crosses, the event may not be detectable at all!

Because the light shadowing has strange effects on the appearance of lunar features, especially the flat maria and craters, I find watching these eclipses very interesting.

If you want to record your observations of a penumbral eclipse, simply note how much of the Moon you think the Earth's penumbra affects. Observe some craters that you have previously drawn and are still familiar with. How do their shadings compare with their appearance when the full Moon is not in penumbral eclipse?

### 21.1.4  Thought

Whether penumbral, partial, or total, lunar eclipses are a pleasant way to mark the place that the Moon passes opposite the Sun, and the beginning of a new 'moondark'. The first lunar eclipse I saw was indeed a partial, although 99.2% of the Moon was covered by the umbra!

The most beautiful lunar eclipse I saw occurred in August, 1989. In order to see it we drove through a thunderstorm, emerging from the clouds a few minutes after totality ended. We saw the Moon also emerging from shadow, framed in a sky full of bright lightning bolts.

## 21.2  Lunar occultations

It is 02:14 hours universal time (UT). Like a hungry monster, the Moon bears down on what appears to be a tiny star, ready to devour it. The distance is rapidly diminishing and there is nothing you can do to stop what is about to happen. Your tape recorder is recording both the steady signal from a standard time radio station such as CHU or WWV, two time signal stations on your short wave set, and your comments on what is happening.

02:14:20: The phone rings. You don't answer it.
02:14:35: You don't want even to blink now, as the star and Moon merge into a single object.
02:14:42.3: Contact! The star has disappeared – just blinked out! To the nearest tenth of a second you have timed an occultation.

Timing occultations, the moment one body passes in front of another, is work that truly is exciting, exacting, and stimulating. Occultations can involve the Moon and a star or planet, a planet and a star, or an asteroid and a star. They can involve satellites of Jupiter or Saturn occulting each other in what we call mutual events. A far more common event occurs when a satellite of a planet like Jupiter transits its planet; these occur quite frequently with Io and Europa, and can be observed and timed using small telescopes. These events are announced each January in *Sky and Telescope* and in the RASC's *Observer's Handbook*. Asteroid occultations are discussed in Chapter 13.

Two methods of recording are popular. One involves a tape recorder that records both the time signals and your calling out 'in' when the star disappears behind the Moon, and 'out' if the event is a star's reappearance. (Unless the star is very bright, we can discern only the events that occur on the Moon's unlit side.) Later, you can listen to the recording and time the event, using a stopwatch, to 1/10 second accuracy.

The other method involves a stopwatch used directly. As soon as the event occurs, you start the stopwatch. Stop it precisely when the first beep starts after the announcer reads 'At the tone, one hour, five minutes . . .' or some such words. Using the occultation form (Figure 21.2) it is easy to deduce the actual observed time of the event.

This field requires a telescope with good optics so that you won't miss the vital moment of disappearance. You also need a pretty good sky and a good idea in advance of whether the Moon will be high enough above the horizon. Even more challenging are the reappearances, during which your idea of the star's relative position to the lunar limb is vital. You need to know where on the limb the star will suddenly reappear. While the bright star egresses and ingresses are always listed in the RASC's *Observer's Handbook*, the International Occultation Timing Association (IOTA) offers expanded lists that include fainter stars.

Amateur observers tend to be attracted to this work because of its exacting nature and because something is actually happening as they look through an eyepiece. Some observers, in fact, have tapes of some of the occultations they have attempted to time, with voices superimposed on radio signals. Their collections of 'WWV's greatest hits' bring back memories of all their occultation experiences.

Occultations teach an observer to be prepared beforehand. The details of the predicted event, including the universal date and time, as well as the position and age of the Moon, are all important so that you can judge whether the Moon will be visible at all from your site the night of the event (I have missed one when the Moon set behind some trees). Make sure also that you have a reliable source for time signals, like a short wave set or some radio that receives one of the time signal stations, and that your tape recorder is working properly with fresh batteries.

In the past, large numbers of occultation observations helped astronomers to determine the Moon's orbit, the precise positions of occulted stars, and the apparent acceleration and retardation of the Moon's speed as it orbited the Earth. These speed changes reflect a somewhat irregular rotation of the Earth. Occasionally occultation timings have clarified our understanding of the shape of a particular feature at the Moon's limb. Also, they have pointed out errors in the catalog positions of certain stars. Finally, occasional

Figure 21.2. Lunar occultation report form, courtesy Royal Astronomical Society of Canada, Montreal Centre.

### MONTREAL CENTRE
### ROYAL ASTRONOMICAL SOCIETY OF CANADA

*Report on Lunar Occultations*

DETAILS OF PREDICTED OCCULTATION

Date (Greenwich) ....................................    (Montreal)...................................

Predicted Time (U.T.)         h.         m.      (E.S.T.)        h.         m.

Z.C. No. ................         Mag. .............         P.A. .............

Phenomenon ...................................         Age of Moon ...................................

DETAILS OF OBSERVATION

Short-wave time signals – Radio Station.................................

                Stopwatch stopped at          h.          m.          s. E.S.T.

                Stopwatch reading          h.          m.          s.*

                Observed time
                of Occultation          h.          m.          s.*

                        * Record time to nearest 1/10 second.

Remarks on Seeing Conditions, etc.:

TELESCOPE USED:

                Type and Aperture............................

                Power of Eyepiece............................

Observer .......................................

Mailing Address ...............................

                ...................................................................

Location of Observing Post...................................................................

                Latitude....................................................

                Longitude....................................................

                Altitude....................................................

detections of anything other than an instantaneous ingress or egress would indicate the presence of an undiscovered binary system.

### 21.2.1 Grazing occultations

Often you might see the Moon approach a star without actually occulting it. It is possible that from a site somewhere else on Earth the Moon actually did occult the star, and from an intermediate place, the Moon 'grazed' the star. When the Moon does not occult a star completely, but instead just grazes it so that the star might pop in and out several times as lunar mountains or valleys pass by, timings of the disappearances and reappearances can provide valuable research information. For grazes, amateur observers mount complex observing programs with observers spaced at various distances from the center of the predicted path. Graze observations still have much scientific value, and teams travel considerable distances to get into their narrow paths. Usually such a network, with portable equipment, has an excellent chance of determining the precise dimensions of various lunar features, and sometimes the Moon itself, when the occultation occurs. Detailed predictions of grazes are available from IOTA.

Grazes are far more complicated to observe than single events, because you are not really sure in advance how many events will take place, the time between them, and the times to expect them. In fact, a few hundred meters distance could decide whether you see no occultation at all, a graze of two to 20 star disappearances and reappearances, or a single occultation event. It is the suspense and excitement, as well as the value to science, that makes graze chasing fun.

### 21.2.2 Occultations of planets

Rarely the Moon will pass in front of one of the major planets. Such an occultation will usually get a considerable amount of advance publicity in the astronomy magazines, as well as in the RASC *Observer's Handbook.* Observe these events just the way you would observe the occultation of a star; you would record the time the planet starts to be occulted, when it is completely occulted, when it begins to emerge, and when the event is completely over. It would also be interesting to note the times when prominent features like Jupiter's red spot are occulted.

### 21.2.3 Occultations by planets

When a major planet is about to occult a star, or apparently pass near one (an appulse), the astronomical magazines will usually announce the event. A planetary occultation is different from a lunar one in that the planet's atmosphere will cause a star to fade over a few seconds, instead of disappearing instantly.

On July 3, 1989, Saturn occulted the sixth magnitude star 28 Sagittarii, and observers across North and South America observed an extraordinary sight. As Saturn's complex ring system passed by, the star faded, disappeared, and reappeared so many times that it was difficult to keep count. As the A ring passed by, I noticed 28 Sagittarii either fade or disappear many times. Then the star resumed its full brilliance as the Cassini division passed by it, followed by a sharp fading as the B ring began its onslaught. The event ended for me with a brief fading that probably signified its occultation by Saturn's slender F ring. The entire performance lasted four hours and was for me the equivalent of watching a performance of Shakespeare. If all that were not excitement enough, the following day European observers observed an occultation of the same star by Saturn's moon Titan!

### 21.2.4 Murphy's Law and occultations

Occultation timing and the proverbial Murphy are appropriate, if frustrating, companions. So much can happen to conspire against you that a successful timing of an occultation can be a thrilling experience, quite beyond your awareness of a new cosmic vista. Courtesy of the Royal Astronomical Society of Canada's Montreal Centre, here are some of the ways that people have 'muffed' an occultation:

Figure 21.3. The Moon about to occult Jupiter. This Steve Edberg photograph was taken on July 16, 1980, through a 25 cm f/7 Newtonian reflector, $\frac{1}{4}$ s exposure, using Ektachrome 400 film.

Figure 21.4. After the occultation, 1/30 s exposure. Since this is a shorter exposure, Jupiter appears smaller. Photo by Steve Edberg.

1. Clouds. Not a totally clouded sky, for that simply means you don't set up, and no time is lost. The cloud to which I refer is a lonely head of vapor that stalks the sky in search of a comfortable place to rest. Just as you have completed setting up, the cloud rushes to the position of the Moon and covers it. After five minutes, both the cloud and your timing have gone.

2. Instrument problems. You set up for an occultation at your friend's house. Too late you remember that all your eyepieces were with your *other* coat.

3. More instrument problems. Your telescope tube feels lighter than usual. You have five minutes to ingress. Too bad! Not enough time to reinstall the mirror you just remembered was out for cleaning.

4. Wind. The tube is shaking so much that you can't tell when ingress or egress occurred. Strong wind can be more frustrating than clouds, as it shakes the telescope and can cause your eyes to water, with blinking or eyepiece fog as an unfortunate result. Even if you see the event, your recorder may pick up the sound of the raging wind so effectively that it loses your voice!

5. Friends. 'What is an occultation?' your (former) best friend asks two seconds before egress.

6. The police. You have moved your telescope out on the street to catch a low Moon just before it sets over your neighbor's house. With stopwatch in one hand and notepad in the other, you clock the occultation just in time to be arrested for voyeurism.

7. A few seconds before ingress, you realize that you forgot to wind your stopwatch or put fresh batteries in your tape recorder, causing your playback to sound like a chipmunk!

8. Checking universal date and time. You have just spent half an hour setting up on the coldest night of the winter to realize that the occultation was *last* night!

# PART SIX

# A miscellany

# 22  Passing the torch

Discovering comets, observing variable stars, timing asteroid occultations . . . which noble astronomical endeavor is the most important thing that an amateur astronomer can do? None of the above. More than any other thing, I believe that the most vital contribution by amateur astronomers is to share their work with the next generation.

Passing the astronomical torch is no trivial matter. It is a difficult, time-consuming task that too often is taken lightly, but with the interest and experience you now have, with credentials like these, you would be the ideal person to go into a classroom. Children appreciate a teacher who loves the subject, and children are very astute at recognizing honest and sincere enthusiasm in a teacher. If you have been interested in astronomy for some time, have tried all the observing branches, children will pick up on your experience and enthusiasm. You chose the stars because you have an affinity for them. You enjoy the peaceful hours by the telescope, no less than the friends and contacts you have made. You have become an observer because you wanted to see the universe for yourself. You may be just the right person to share this universe.

What if you are a beginner, reading this to get some motivation to enjoy the stars? What better way is there to enjoy the sky than with a child – or a parent – who can discover the stars with you?

## 22.0.1  Schools

The reason this activity is so important is that if we, who have been struck by the astrobug, do not transmit our insights to a new generation, there will be no next generation in astronomy. All but ignored throughout North America's public schools, astronomy averages less than one hour per day for a few days each year. Generally the sky is introduced by teachers who know very little about the subject themselves. It would be so helpful if some of these hours could be augmented by someone with the knowledge and enthusiasm to describe the stars to children. An astronomy club could provide invaluable help in arranging these special sessions.

If you have joined the astronomy club, find out what its children's programs are, and if there aren't any, encourage it to get involved with kids. Not enough clubs realize the potential that working with children has. When working with young people, we use everything we have learned in observing, as well as all the skills we can generate as communicators. Children are difficult but worthwhile audiences, and if you succeed in arousing their interest, you will have accomplished something important.

## 22.1  Methods of teaching

Whether in schools, summer camps, or informal groups, the biggest mistake I have made in my own teaching is to underestimate the intelligence or the imagination of a child. Of all the discoveries I have made in my observing years, one of the most satisfying is the incredible ability of a young child to grasp the basic concepts of the sky. The important thing is to find some children and point their way to the stars.

Figure 22.1. Either end is fine!

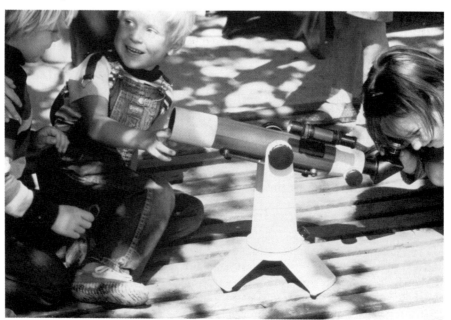

## 22.2   The planets

Following the Socratic method, ask questions to find out what the children already know, and to encourage them to learn things for themselves. Expect to be surprised; most children know the names of the nine major planets, and even a bit about the characteristics of each. The planets are fun to talk to kids about since they are real places with atmospheres and environments that they can relate to.

Most of the children I have taught prefer that mysterious planet Mars. As a planet with character and history (both there and here on Earth), Mars has a proven ability to inspire children. Possibly the reason for the inspiration is the red color; all you have to do is say 'red planet' and their minds become inquisitive: 'Why red? All the stars are white!'

Then we get into a little history. Schiaparelli thought he saw channels, and in 1896 Lowell's observations of the 'canals' created even more attention. Years later, Orson Welles narrated a radio play called *War of the Worlds*, terrifying thousands of listeners who, glued to their radios, really did believe that the world was coming to an end.

I mimic the program, taking both sides in the conflict between Martians and Earthlings in an attempt to bring the old fictitious battle to life. At the height of the fighting, I quickly change gears and ask about Mars itself. Would the Martians be comfortable on this planet, or were they fighting us because of our lack of air conditioning? After all, creatures comfortable at subzero temperatures would be devastated by our room temperature.

Let's continue this game with the other planets, and with more than just the weather. Would you like to live on Venus? That one is hard to imagine, unless we have a visitor from that hot planet. Crawling through our window, the Venusian shivers his way towards our oven, which is conveniently set at 450 degrees Fahrenheit to cook a frozen dinner. Crawling into the oven, the hapless Venusian nevertheless freezes to death. Freezes to death in an oven set to 450? Think of the children's expressions when they learn that the oven would have to be twice as hot to handle the comfort level of the Venusian!

In this personal way, we can either travel to the other planets, boiling on Mercury, dying from overweight, poison gas, and radiation above Jupiter, or freezing on Pluto. This is how we acquaint the planets to children, relating distant concepts to real lives.

## 22.3   Daytime observing

Up to now our discussion, however, is still theoretical, and in a book that is encouraging observation, we must come up with some observing sessions for the children. This is a difficult process. In a world where our contacts with kids are usually limited to school hours, how can we get the children out observing?

### 22.3.1  Observing the Sun

The most tempting and obvious daytime object is the Sun. With a little advance precaution, you can impress safety on the children. Using the focused rays of the Sun and a plastic garbage bag at the focal point, ask the children to guess how long it will be before the bag ignites, and observe their shock when smoke appears before you have even finished your question! Imagining the same thing happening to their eyes, which you explain are even more delicate than the bag, keeps the kids in line.

Using a projection system (see Chapter 8) allows you to show the Sun to a large group in a very short time, and lets you repeat the experiment day after day to show how the Sun changes.

### 22.3.2  Venus

We can sometimes return the visit of our Venusian friend; if Venus is in the nighttime sky, then it is easy to find during daylight; see Chapter 2 for instructions.

Let the kids enjoy what may be their first look through a telescope, and almost certainly their first telescopic introduction to the planets, giving each child enough time to enjoy the view. Make sure the children are looking at the right thing by having them report to you what they see. Sometimes they may need extra time before their eyes 'know what they are seeing' and Venus pops into their frame of reference.

Figure 22.2. Projecting the Sun into a box.

### 22.3.3 Observing the Moon

An observing session in daylight could track the Moon with ease, and, like Venus, all you really need is an idea of where the Moon is in the sky. If the Moon is near its full or new phases, it will not be easily visible in daylight. As the Moon passes its first quarter, it is easy to see during the afternoon, and makes a fine subject for a telescopic look. Around last quarter the Moon rises during the night, and is still in the sky during the morning hours.

The bright daytime sky reduces contrast between the Moon and the sky. Obviously for a beginner, the night view is far superior. When showing the daytime Moon to children, emphasize that at night the same Moon would look much more dramatic. Encourage them to use binoculars or a telescope, if they have one, to carry on their learning past classroom hours.

## 22.4 Night observing

If you are fortunate enough to get children out under the stars at night, point out some constellations. Then ask the children to study the sky and invent their own constellations. Is there a dog or cat in the sky? By forming their own imaginary figures, they create their own sky.

When the telescope time arrives (usually after about half an hour of

Figure 22.3. No telescope is too big for kids.

observing) try something bright like the Moon. By bending down on the opposite side of your telescope, you can look at the child's observing eye to ensure that moonlight actually is entering it. Be patient so that each child gets as much time as necessary for this critical first look.

If the Moon is not up, try one of the planets. Jupiter, Saturn, and Venus are good choices; Jupiter's four big satellites and two central cloud bands, Venus's phase, and Saturn's rings are visible even through a small refractor. These planets offer a special bonus for observers when they are high in the sky. Easy to find, they are bright enough to catch the children's attention first without a telescope, so that they know what they are about to look at.

What if no major solar system object is visible? I would then suggest a double star like Gamma Leonis in spring, Albireo in summer, Gamma Andromedae in autumn, or Theta Orionis in the Orion Nebula in winter. Also, Mizar, the middle star in the handle of the Big Dipper, is visible almost the year round from north temperate latitudes.

## 22.5  Meteors, and learning through research

If children can like stars, they'll love meteors. Offering something in the sky that changes, meteors help bring the sky to life. If you happen to have some children around near the peak of a meteor shower, why not organize

Figure 22.4. Is Rachael old enough to meet a telescope?

them into an observing team (see Chapter 3) and have them count meteors for an hour or two?

During the Perseid maximum in 1978, I conducted a watch for a group of children aged six to nine. Whenever a child reported a meteor, I would ask questions about the path, the brightness, and any special effects the child might have seen. We ended the watch with statistics that seemed reasonable and, with appropriate comments, were reported to a national meteor group.

Older children can make quite proficient meteor observers. In 1966

Figure 22.5. All photos in this chapter are by the author.

groups observing the Delta Aquarids reported 685 and 784 meteors on two nights around maximum. The watches lasted for several hours each night, the children had no trouble staying awake, and the nights were laced with the humor and good spirits that mark an enthusiastic group. At the time, Canada's National Research Council sponsored a visual meteor program, and the results were proudly reported to it.

When programs like this result in data that are reported, the children are learning through active research. Although their results will likely not lead to anything substantially new, they show youngsters that science welcomes their work, and that should inspire them to go on. A few high school teachers have indulged this theme even further, creating advanced projects for high schoolers to work on. Occasionally this may even include getting time at an observatory telescope. Programs like this are rare and offer children a special chance at astronomical success, but their teachers should not conduct such programs at the expense of the basic observing, including observing constellations and drawing planets, that is so important.

## 22.6    Closing thought

Some of the most valuable observing sessions I have are with children. Their questions are intriguing. Who else but a ten-year-old would look through your telescope at Jupiter and say, 'Why were we put here, and not there?' Now that you have found your way into astronomy, here is your chance to pass your interest on to the heart and mind of someone younger. It is a time to reflect on the 1859 words of Senator Carl Shurz:

> Ideals are like stars; you will not succeed in touching them with your hands. But like the seafaring man on the desert of waters, you choose them as your guides, and following them you will reach your destiny.

# 23    The poet's sky

Astronomers are poets, Leslie Peltier once said to Walter Scott Houston. In a casual conversation between two of the century's most highly respected amateur astronomers, the concept of observers and poets was discussed. Perhaps the idea of poetry helps define the amateur tradition in astronomy; we see what poets see when we look upward.

Poets have gazed skyward as a symbol of lofty aims at least since Quintus Ennius wrote that 'No one regards what is before his feet; we all gaze at the stars.'[1] In 1579 Edmund Spenser's 'Shepheardes Calender' continued that thought:

> He that strives to touch the stars
> Oft stumbles at a straw.

1 Cicero, *De Divinatione* II, ch. 13.

Three centuries later, in 'Lady Windermere's Fan' (1892), Oscar Wilde gave the idea another turn; 'We are all in the gutter, but some of us are looking at the stars.'

Poets have responded in other ways to the stars. Geoffrey Chaucer was quite capable of discussing scientific instruments seriously, as we can see in his 'Treatise on the Astrolabe'. However, he could have fun with astronomers as well. In words written around 1390, this passage from 'The Miller's Tale' virtually foretold comet hunter Charles Messier's fall into a wine pit while looking for comets (see Chapter 14) almost four centuries later:[2]

> He walked in the feeldes, for to prye
> Upon the sterres, what ther sholde bifalle,
> Till he was in a marle-pit yfalle;
> He saugh nat that.
>
> (lines 3458–61)

In those days astrology was taken seriously, thus Chaucer's depiction of the astronomer wondering what the stars will bring really does reflect what astronomers did. As late as Shakespeare's time, the word 'astronomy' involved what we now regard as astrology.

Just before his death in 1543, Nicholas Copernicus published his theory about a heliocentric universe. Since the ideas of Copernicus were theoretical, poets took little notice of them. Shakespeare's allusions are almost entirely limited to the astrological effects of the stars. As Ophelia reads a letter from Hamlet, the reference to Copernicus is meant to be taken lightly:[3]

> Doubt that the stars are fire,
> Doubt that the sun doth move,
> Doubt truth to be a liar,
> But never doubt I love.
> (*Hamlet*, Act II, scene ii, lines 116–19)

Only a few years later, Galileo's observations demonstrated that a heliocentric system might be correct. Poets could not ignore the new evidence. Only one year after Galileo published his early observations, John Donne commented on the new philosophy in 'Anatomie of the World'. Where Shakespeare may have been having fun, Donne was writing in a very serious vein:[4]

2 Geoffrey Chaucer, *The Works of Geoffrey Chaucer*, ed. F.N. Robinson, p. 51. Boston: Houghton Mifflin, 1957.

3 W. Shakespeare, *Hamlet*, II, ii, 116–19.

4 John Donne, *The Poems of John Donne*, ed. Sir Herbert Grierson. Oxford University Press, 1933.

And new philosophie calls all in doubt,
The Element of fire is quite put out;
The Sun is lost, and th'earth, and no mans wit
Can well direct him where to look for it.
And freely men confesse that this world's spent,
When in the Planets, and the Firmament
They seek so many new; they see that this
Is crumbled out againe to his Atomies.
'Tis all in pieces, all coherence gone;
All just supply and all Relation:
Prince, Subject, Father, Sonne, are things forgot,
For every man alone thinkes he hath got
To be a Phoenix, and that then can bee
None of that kinde, of which he is, but hee.

(lines 205–18)

This poem displays a great feeling of distress by the loss of the old system and does not show much confidence in what Galileo's telescope has found to replace it. A decade later, in 1622, Sir John Davies intended his 'Orchestra' to be a celebration of the cosmic order. Fearing that what the telescope would reveal would upset that order, he included this caveat:[5]

Onely the Earth doth stand for ever still,
Her rocks remove not, nor her mountaines meete,
(Although some witts enrich with Learnings skill
Say heav'n stands firme, and that the Earth doth fleete
And swiftly turneth underneath their feete)
Yet though the Earth is ever stedfast seen,
On her broad breast hath Dauncing ever been.

(stanza 51, lines 103–4)

The poetic debate about the new order was not resolved even by the publication of *Paradise Lost* in 1667. Wanting to write an epic for all time, Milton hedged his bets, making sure that the angel Raphael did not commit himself to a geocentric or a heliocentric universe when he answered, in Book VIII, this question posed by Adam:[6]

What if the Sun
Be Centre to the World, and other Stars
By his attractive virtue and their own
Incited, dance about him various rounds?
Thir wandring course now high, now low, then hid,
Progressive, retrograde, or standing still,
In six thou seest, and what if sev'nth to these
The Planet Earth, so steadfast though she seem,
Insensibly three different motions move?

(lines 122–30)

5 Sir John Davies, *The Poems of Sir John Davies*, ed. R. Krueger. Oxford: Clarendon Press, 1975.
6 John Milton, *Complete Poems and Major Prose*, ed. M. Hughes, p. 365. Indianapolis: Odyssey Press, 1957.

After Herschel discovered Uranus in 1781, telescopes became more popular and widely used. In his 1806 poem 'Star Gazers' William Wordsworth alluded to the disappointed feeling in a group of people lined up to peer through a telescope:[7]

> Whatever be the cause, 'tis sure that they who pry and pore
> Seem to meet with little gain, seem less happy than before:
> One after one they take their turn, nor have I one espied
> That doth not slackly go away, as if dissatisfied.

Have you had this experience? Looking through a telescope for the first time, you expected to see something like the pictures or drawings made with large telescopes. To turn your initial 'dissatisfied' reaction into an appreciation of what a telescope has to offer was a central aim of this book.

The discovery of Neptune in 1846 occurred only four years before the publication of Tennyson's great epic *In Memoriam*. I think that this poem's comment alludes to that find:[8]

> A time to sicken and to swoon
>> When Science reaches forth her arms
>> To feel from world to world, and charms
> Her secret from the latest moon?
>> (XXI, lines 17–20)

One can travel either way on the road that joins astronomy with poetry, as an observer with poetry or a poet who observes. Gerard Manley Hopkins, one of the nineteenth century's best known English poets, was both, and it is possible that he observed the clear, moonless pre-dawn sky of August 4, 1864. Although both Jupiter and Saturn had set, Mars was enjoying a favorable opposition (see Chapter 11) and was well placed in the southern sky. The constellations of Auriga and Taurus were prominent, and Gemini was rising in the northeast. Just west of the second magnitude star Beta Tauri shone the head of Tempel's Comet, and its tail stretched towards nearby Iota Aurigae.

Tempel's Comet was moderately bright for only a short time as it sped near the Earth. On September 13 the young poet wrote these lines:[9]

> — I am like a slip of comet,
> Scarce worth discovery, in some corner seen
> Bridging the slender difference of two stars,

7 William Wordsworth, *The Poetical Works of William Wordsworth*, ed. T. Hutchinson, p. 189. Oxford University Press, 1917.

8 Alfred Lord Tennyson, *Victorian Poetry and Poetics*, eds. W. Houghton and G. Stange, p. 51. Boston: Houghton Mifflin, 1967.

9 Gerard Manley Hopkins, *Poems of Gerard Manley Hopkins*, 4th edn., eds. W.H. Gardner and N.H. MacKenzie, p. 147. Oxford University Press, 1970.

Come out of space, or suddenly engender'd
By heady elements, for no man knows:
But when she sights the sun she grows and sizes
And spins her skirts out, while her central star
Shakes its cocooning mists; and so she comes
To fields of light; millions of travelling rays
Pierce her; she hangs upon the flame-cased sun,
And sucks the light as full as Gideon's fleece:
But then her tether calls her; she falls off,
And as she dwindles sheds her smock of gold
Amidst the sistering planets, till she comes
To single Saturn, last and solitary;
And then goes out into the cavernous dark.
So I go out: my little sweet is done:
I have drawn heat from this contagious sun:
To not ungentle death now forth I run.

Hopkins is describing a comet in human terms. The comet is first seen 'bridging the slender difference of two stars'. While doing research for my 1979 Master's thesis at Queen's University, I found a letter to *The Times* (of London) of August 1, 1864, which predicted the passage of this comet 'between the stars Iota in Auriga and Beta in Taurus'.

The poet's question whether comets come from space or are born in our atmosphere, and the comment about Saturn being the last planet, are not out of place; Hopkins had intended this piece as a part of a play that was set in Italy during the Renaissance. Hopkins intended the poem to reflect the understanding of our solar neighborhood that prevailed at that earlier time.

The centers of some bright comets may indeed shake their cocooning mists, as Hopkins suggests. With each rotation, the nucleus of a major comet turns its more active side to the Sun, resulting in periodic eruptions of material. Because Halley's Comet shedded its cocooning mists in the form of major jet eruptions every 2.2 days, these were successfully predicted by the International Halley Watch Near Nucleus Studies Network in time for use during the spacecraft encounters. If Hopkins could have been here to see this he may not have been astonished. He probably suspected that the regular emanations of 'cocooning mists' from the nucleus were important to our understanding of these celestial visitors, and therefore wrote about them in his poem.

One need not discover the poetry that is so rich in astronomy only through observing. Through the lenses he ground, John Brashear knew the joy of turning a piece of raw glass into a magic window to the heavens. Over his grave and that of his wife Emily are words that are also in the minds of many of us: 'We have loved the stars too fondly to be fearful of the night.'

Our fondness for the stars has touched our souls. We all share the feeling of discovery, whether the object we have found is new to all or new only to us. The thrill penetrates our being, and we try to describe through drawings, photographs, or words, how we have been changed by the universe sharing a secret with us. It is a feeling that observers and poets experience. Walt Whitman's 'When I heard the learn'd Astronomer' takes to task the lecturer who recited 'the facts and figures' leaving the listener 'tired and sick' – he left the room and

> In the mystical
> moist night air, Looked up in perfect silence at the stars.

In silence or not, only by looking at the stars can we really appreciate what they have to offer. Only by looking can we share the thrill that the poet Ralph Hodgson felt in 1913 at the conclusion of his 'Song of Honour':

> I stood and stared; the sky was lit,
> The sky was stars all over it,
> I stood, I knew not why,
> Without a wish, without a will,
> I stood upon that silent hill
> And stared into the sky until
> My eyes were blind with stars and still
> I stared into the sky.

# 24   Resources

## 24.1  Societies

If you want a first hand look at the sky, astronomy clubs are for you. Many of the hints and suggestions I have made about the kind of telescope you should buy come to life at a club star party, where you actually get to use all these different telescopes. Join a club and you get to test many types and brands of telescopes and binoculars before you buy them. Astronomy clubs and national organizations provide your doorway to meeting others who share your interests, to providing you with access to updated information, and to contributing observations that would increase your enjoyment of the stars and possibly make a contribution. They provide one of our most enduring resources.

In that light alone, joining a club is a good investment. But clubs offer much more. Most clubs have good monthly meeting programs, during which you can hear speakers tell you about different aspects of astronomy and sister sciences. They also offer newsletters, and often discount subscriptions to *Sky and Telescope* or *Astronomy* magazines. Some clubs give their members the added benefit of contacts with amateur astronomers across the country and the world, by affiliating with national societies. They also can have annual banquets or special meetings during which their members can learn and be inspired from internationally known people in various fields of astronomy.

The two disadvantages of amateur clubs that I have seen involve politics and the program level. Some organizations tend to get too political. I guess that politics is an unavoidable consequence of people getting together, although some clubs may carry it too far. If old business and executive meetings are not your cup of tea, you may find this part of a club uninteresting, but one hopes that clubs would keep much of the political minutiae to executive meetings. Also, some clubs set their program level too far above that of a beginner. If you can't understand what someone is saying, don't be afraid to ask. 'My question is I don't understand it' is OK, and might be the only way to learn!

A description of some national organizations follows. If you are interested in the work they do, you may write to them at the addresses I have included. (Of course it is possible that addresses will become outdated over time.) The best way to become more familiar with a particular branch of astronomy is to join a society that specializes in it. The groups listed below cater both to beginning and advanced observers, and they do offer training programs to encourage newcomers. Sometimes you may find that you have to write two or three times before you get a response. While this is unfortunate, do remember that these groups are run either by volunteers or on very tight budgets, and that generally they try to answer queries and certainly are anxious to spread enthusiasm for their fields. Including a stamped, self-addressed envelope is a courteous thing to do.

**Lunar and planetary** Association of Lunar and Planetary Observers (ALPO). John E. Westfall, Director; PO Box 16131, San Francisco, California, 94116, USA. In 1947, Walter Haas, an experienced and prominent amateur astronomer from New Mexico, founded this organization to encourage solar system observation. Since then, this group has spearheaded amateur contributions to the observation of the Moon and planets. Several of today's most prominent planetary scientists had their career beginnings with this special group. The ALPO emphasizes visual observing and photography of the Moon, the planets, and the Sun. Their work is divided into sections by body, with at least one recorder for each.

**Variable stars** American Association of Variable Star Observers (AAVSO). Janet Mattei, Director; 25 Birch Street, Cambridge, Massachusetts, 02138, USA. Since 1911 this group has encouraged the systematic observation of variable stars. This is the world's largest organization devoted specifically to the encouragement and archiving of observations of variable stars. The association has two meetings per year, one in northern hemisphere spring and the other in autumn. Although its main organization considers most variable star observations, there are special sections for sunspot counting, eclipsing binary work, Cepheid observing, and nova and supernova searches. The British Astronomical Association's Variable Star Section also does very useful variable star work. The Royal Astronomical Society of New Zealand's Variable Star Section, PO Box 3093, Greerton Tauranga, New Zealand, has done much to increase our understanding of many southern hemisphere variable stars.

**Occultations** International Occultation Timing Association (IOTA). David Dunham, Director; 1177 Collins, Topeka, Kansas, 66604, USA. This group devotes itself entirely to the systematic observation and timing of occultations by the Moon, asteroids, or planets, and of solar eclipse contacts.

**Photometry** International Amateur–Professional Photoelectric Photometry (IAPPP). Fairborn Observatory, Dyer Observatory, Vanderbilt University, Nashville, Tennessee, 37235, USA. The IAPPP seeks to bring advanced amateurs and professionals together in the field of photoelectric photometry, which measures the light output of stars using either light measuring devices called photometers or charge-coupled devices (CCDs).

**Deep sky** National Deep Sky Observers Society (NDSOS). Alan Goldstein, Director; 3430 Bryan Way, Louisville, Kentucky 40220, USA. A new group designed to foster the observation of deep sky objects, it has special 'observing clubs' that devote themselves to specific types or levels of difficulty of deep sky objects. The Webb Society, 1 Hillside Villas, Station Road, Pluckley, Tenterden, Kent, TN27 Q2X, UK, is an international organization devoted to the study of double stars, clusters, nebulae, and galaxies.

**Light pollution** International Dark Sky Association. 3545 North Stewart Avenue, Tucson, Arizona, 85716, USA. This organization was founded in 1987 by David Crawford, an outstanding professional astronomer at the National Optical Astronomy Observatories, and Tim Hunter, an amateur astronomer then President of the Tucson Amateur Astronomy Association, to bring attention to the major problem of light pollution over our cities. Through a newsletter and an annual meeting, this group is trying to point out ways to light cities safely and economically in such a way that the sky above will be also well preserved.

**Sun** Solar Section, ALPO. Richard Hill, Recorder; 4632 E. 14th Street, Tucson, Arizona, 85711, USA.

Solar Section, AAVSO. 25 Birch Street, Cambridge, Massachusetts, 02138, USA.

These two groups encourage observation of the Sun. While the AAVSO group monitors the numbers of sunspot activity, the ALPO section is more concerned with morphology of sunspot groups.

**Four other organizations** The Royal Astronomical Society of Canada (RASC), 124 Merton Street, Toronto, Ontario M4S 2Z2, Canada. One of the world's largest 'general-purpose' astronomical societies, the RASC seeks to bring Canadian astronomers together. Unlike most amateur organizations, this society has a substantial number of professional astronomers in its membership. Divided into 'centres' based in cities across the country, which hold regular meetings and publish newsletters, the national organization conducts an annual 'General Assembly', publishes a professional Journal and a less formal *National Newsletter*, and the famous annual *Observer's Handbook*.

The Astronomical League, Science Service Building, 1719 N. Street, NW, Washington, DC, 20036, USA. Many of the amateur societies in the United States are members of the Astronomical League, a federation of societies. It offers a quarterly newsletter called *The Reflector*.

Western Amateur Astronomers is a federation of amateur clubs based in California and other western states. Their main purpose is the holding of an annual meeting that brings together amateurs from all over the western USA.

The Astronomical Society of the Pacific, 390 Ashton Avenue, San Francisco, California, 94112, USA. Based in San Francisco, this worldwide organization has the triple purpose of serving its professional members through a Journal, and its amateur members through a magazine called *Mercury* and teachers with a newsletter called *The Universe in the Classroom*. An annual conference joins all three groups.

## 24.2  Literature

There are hundreds of books available on telescopes and astronomy, and the few I have selected are among the best. This list does not include all the worthwhile books, but it is intended as a guide for further reading.

### 24.2.1  Observing assistance

Berry, Richard. *Discover the Stars*. New York: Harmony Books, 1987. *Written by the editor of* Astronomy *magazine,* Discover *will bring you to the level of the book you are reading right now. Star maps, clearly described and arranged month by month, make it easy for you to learn the sky on your own schedule. Constellations are described in a lively way that makes the book a good starting point.*

Burnham, Robert. *The Star Book*. Milwaukee: AstroMedia, and Cambridge University Press, 1987. *The monthly maps in this guide are more detailed than those in* Discover. *Sturdy, spiral-bound book designed for use in the field.*

Burnham, Robert, Jr. *Burnham's Celestial Handbook*. New York: Dover, 1978. *Very detailed three-volume collection of descriptions of stars, double stars, variable stars, deep sky objects, as well as general background information, all arranged by constellation.*

Couteau, Paul. *Observing Visual Double Stars*, trans. Alan Batten. Cambridge, MA.: MIT Press, 1982. *Thorough treatment of double stars.*

Dobbins, Thomas A., Parker, Donald C., and Capen., Charles F. *Introduction to Observing and Photographing the Solar System*. Richmond: Willmann-Bell, 1988. *A good guide to observing the planets and other solar system objects except the Sun. The chapters on photographic observing are particularly good.*

Edberg, Stephen J., and Levy, David H. *Observe Comets*. Washington: Astronomical League, 1985. *An observing manual for comet observers.*

Eicher, David. *The Universe From Your Backyard*. Milwaukee: Kalmbach Publishing, 1988. *A collection of 'Backyard Astronomer' columns from* Astronomy *magazine. As a book, the collection guides you through a tour of deep sky objects that you can see from your own back yard.*

Kukarkin, B. V., *et al*. *General Catalogue of Variable Stars*, 4th edn. Moscow: Astronomical Council of the Academy of Sciences in the USSR, 1985. *If you are interested in a complete listing of all known variable stars, this valuable research source lists every known variable star at the time of its publication.*

Levy, David H. *The Joy of Gazing*, 2nd edn. Montreal Centre of the Royal Astronomical Society of Canada, 1985. *Actually the seed from which the book you are now reading grew, this small booklet presents ways of getting started in several areas of observing.*

Levy, David H. *Observing Variable Stars: A guide for the beginner*. Cambridge University Press, 1989. *Great book, fine publisher.*

Levy, David H., and Edberg, Stephen J. *Observe Meteors*. Washington: Astronomical League, 1986. *Guide for people interested in observing meteors.*

Mayer, Ben. *Starwatch*. New York: Perigee, 1984. *This book is designed for use with 'Starframes' – coathangers with plastic wrap, and painted stars – that you then look through to find different constellations. An original and fun way to learn the constellations.*

Muirden, James, ed. *Practical Amateur Astronomy*. Essex: Longmans, 1991. *An observing guide designed for a moderately advanced amateur audience.*

Peltier, Leslie C. *Guideposts to the Stars*. New York: Macmillan, 1972. *Peltier's book reaches out to advanced elementary and early secondary school children to point out the stars and the constellations clearly.*

Peltier, Leslie C. *Leslie Peltier's Guide to the Stars: Exploring the sky with binoculars*. Milwaukee: AstroMedia, and Cambridge University Press, 1986. *Excellent guide, in which each season's binocular highlights are clearly described, along with some hints to more advanced binocular observing in several areas.*

Salmi, Juhani. *Check a Possible Supernova*. Vesijarvenk, 36 C 40, 15140 Lahti, Finland, 1984, 1985. *This astrophotographer has published two sets of galaxy photographs. Spiral bound, they provide a convenient reference source for supernova hunters.*

Sidgwick, J. B. *Introducing Astronomy*, rev. edn. London: Faber, 1973. *A reprint of a classic, this book has a very long appendix in which each constellation is drawn and described along with the interesting things to see.*

Sidgwick, J. B. *Observational Astronomy for Amateurs*, 4th edn, revised by J. Muirden, Hillside, N.J.: Enslow Publishers, 1982. *Excellent section on methods for observing variable stars.*

Sinnott, Roger W., ed. *NGC 2000.0* Cambridge, Mass.: Sky Publishing Corporation, 1988. *Complete list of the* New General *and* Index Catalogues, *precessed to equinox 2000.0, so that the positions in this book match any star atlas precessed to the equinox of the year 2000.*

### 24.2.2  Annual guides

Bishop, Roy, ed. *The Observer's Handbook.* Toronto: The Royal Astronomical Society of Canada. *Superb annual guide to positions of the planets, times for eclipses, and to what special events can be seen from night to night.*

Ottewell, Guy, ed. *Astronomical Calendar. Coffee table format guide to many areas of observing, including the sky each month, comets, meteors, asteroids, and space flight.*

Westfall, John E., ed. *The ALPO Solar System Ephemeris.* San Francisco: Association of Lunar and Planetary Observers. *Much useful information for planning observation of solar system objects.*

### 24.2.3  Star atlases

Dickinson, Terence, *et al. Mag 6 Star Atlas.* Barrington: Edmund Scientific, 1982. *Large format, clearly written and explained. The lists of special objects for each chart are original and helpful.*

Norton, Arthur P., and Inglis, J. *Norton's 2000.0*, 18th ed., I. Ridpath. Cambridge, Mass.: Sky Publishing Corporation, 1989. *Contains 16 charts followed by a long introductory section. Earlier editions also contain the lunar map with the 326 features that we discussed in Chapter 6.*

Scovil, Charles E. *The AAVSO Variable Star Atlas.* Cambridge, Mass.: Sky Publishing Corporation, 1980. *Excellent guide for finding variable stars. Also useful for finding other deep sky objects.*

Tirion, Wil. *Sky Atlas 2000.0.* Cambridge, Mass.: Sky Publishing Corporation, and Cambridge University Press, 1981. *Very good atlas of the entire sky to magnitude 8.0. Includes clusters, nebulae, and galaxies as well. This is one of the most popular telescopic star atlases available.*

Tirion, Wil, Rappaport, Barry, and Lovi, George, *Uranometria 2000.0.* Richmond: Willmann-Bell, 1987, 1988. *Title based on Bayer's* Uranometria *of 1603. For more experienced telescope users. Two volumes contain 473 charts covering the entire sky.*

### 24.2.4  Historical

Allen, Richard Hinckley. *Star Names: Their lore and meaning.* G. E. Stechert, 1899. Reprint, New York: Dover, 1963. *A classic collection of mythological and historical stories about the names of the constellations and hundreds of individual stars.*

Ashbrook, Joseph. *The Astronomical Scrapbook*. Cambridge, Mass.: Sky Publishing Corporation and Cambridge University Press, 1984. *A compilation of years of* Sky and Telescope's *'Astronomical Scrapbook' columns, a treasure of astronomical stories of both scientific and human interest.*

Bronowsky, J. *The Ascent of Man*. Boston: Little, Brown, 1973. *The written version of the magnificent television series. The chapter called 'The starry messenger' provides useful background to the Galileo story.*

Peltier, Leslie C. *Starlight Nights: The adventures of a star gazer*. Harper & Row, 1965. Reprint, Cambridge, Mass.: Sky Publishing Corporation, 1980. *Exquisite book; the story of the life of a man who had been observing tne night sky for over 60 years. No other book captures so well the basic feeling of what it is like to be an observer.*

Preston, Richard. *First Light: The search for the edge of the Universe*. New York: New American Library, 1988. *A wonderful adventure story, including little known insights about the lives of astronomers past and present, and generally related to Palomar Mountain Observatory. A section over 60 pages long discusses the work of comet discoverers Eugene and Carolyn Shoemaker at Palomar Observatory. Stories in this book reminisce about George Ellery Hale and the building of Palomar, Maarten Schmidt's discovery of the remoteness of quasars in 1963, Bernhard Schmidt and the development of the Schmidt camera, and other fascinating events.*

Tombaugh, Clyde W., and Moore, Patrick. *Out of the Darkness: The planet Pluto*. Harrisburg, Pa.: Stackpole Books, 1980. *A book about the discovery of Pluto, co-written by the man who found this distant planet half a century ago. Very exciting reading.*

## 24.2.5 Solar system

Beatty, J. Kelly, and Chaikin, Andrew, eds. *The New Solar System*, 3rd edn. Cambridge, Mass.: Sky Publishing Corporation 1990. *If you want an up-to-date compendium of what we understand about the solar system, including Voyager summaries from Jupiter, Saturn, Uranus, and Neptune, this is the best source.*

Chapman, Clark R. *Planets of Rock and Ice: From Mercury to the moons of Saturn*. New York: Scribner's, 1982. *The best of the summaries of what we now know about these planets. Easy to read; concepts are presented so that they are easily understood.*

Chapman, Clark, and Morrison, David. *Cosmic Catastrophes*. New York: Plenum, 1989. *A fine book describing how dangerous our immediate space environment is, with facts on past encounters with comets and asteroids, and ideas about future hits.*

Cunningham, Clifford. *Introduction to Asteroids*. Richmond: Willmann-Bell, 1988. *Thorough, basic introduction.*

Whipple, Fred L. *The Mystery of Comets*. Washington: Smithsonian Institution Press, 1985. *Written by the most noted student of comets of the 20th century, this book contains a thorough discussion of comet history and science.*

## 24.2.6  Deep sky

Bok, Bart J., and Bok, Priscilla F. *The Milky Way*, 5th edn. Cambridge, Mass.: Harvard, 1981. *Since I had the privilege of knowing the first author so well, I find it hard to be objective about a book that belongs in every thinking person's library. Go out and buy a copy while you still can, and I hope you enjoy it as much as I did. A classic.*

Eicher, David, ed. *Deep Sky Observing with Small Telescopes*. Hillside, N.J.: Enslow, 1989. *Chapters by Glenn Chaple on double stars, David Levy on variable stars, globular clusters, and bright and dark nebulae, David Eicher on open clusters, Michael Witkoski on planetary nebulae, and Alan Goldstein on galaxies.*

Mallas, John H., and Kreimer, Evered. *The Messier Album*. Cambridge, Mass.: Sky Publishing Corporation, 1978. *Each Messier object is described with basic data and personal observations. Clearly written, valuable introduction to the Messier objects.*

Newton, Jack. *Deep Sky Objects: A photographic guide for the amateur*. Toronto: Gall Publications, 1977. *Includes photographs of most of the Messier objects. Useful guide for finding Messiers and some other deep sky objects.*

Newton, Jack, and Teece, Philip. *The Cambridge Deep-Sky Album*. Cambridge University Press, 1983. *A collection of amateur photographs of Messier and other objects. The galaxy photos are useful sources for searchers of supernovae.*

## 24.2.7  General assistance

Berry, Richard. *Build Your Own Telescope*. New York: Charles Scribner's Sons, 1985. *A complete and well-written guide to building simple telescopes.*

Dickinson, Terence. *Night Watch*. Camden East, Ontario: Camden House, 1983; revised and extensively updated, 1989. *This book combines two different aspects of astronomy for beginners; a subject that can be studied, and a hobby to be enjoyed. Written by an author with experience in both areas of the field.*

Dickinson, Terence. *The Universe and Beyond*. Camden East, Ontario: Camden House, 1986. *Beautifully illustrated with photographs and art work, this well-presented introduction is a joy to page through.*

Ferris, Timothy. *The Red Limit: The search for the edge of the Universe*. New York: Morrow, 1977. *An exciting cosmological read.*

Fjermedal, Grant. *New Horizons in Amateur Astronomy*. New York: Perigee, 1989. *A book about amateur astronomy, based extensively on conversations with active amateurs.*

Hogg, Helen Sawyer. *The Stars Belong to Everyone*. Toronto: Doubleday,

1976. *A landmark. This Canadian professional astronomer has written a superb introduction to astronomy.*

Moore, Patrick. *The International Encyclopedia of Astronomy.* New York: Orion Books, 1987. *Good alphabetic coverage of astronomical topics, supplemented by some major essays on selected topics.*

Sagan, Carl. *The Cosmic Connection.* New York: Dell, 1973. *Penetrating and incisive comments about our place in the universe.*

Sagan, Carl. *Cosmos.* New York: Random House, 1980. *A printed memory from the outstanding television series.*

## 24.2.8  For children

Chaple, Glenn F., Jr. *Exploring With A Telescope.* New York: Franklin Watts (Venture), 1988. *Excellent introduction to telescopic observing for children.*

Dickinson, Terence. *Exploring the Night Sky: The equinox astronomy guide for beginners.* Camden East, Ontario: Camden House, 1987. *Worthy guide for children. Takes young explorers on a ten-step cosmic voyage in which different aspects of the universe are explored.*

Levy, David H. *The Universe for Children.* Oakland, Ca.: Everything in the Universe, 1985. *How astronomy-minded adults can help children to love the sky.*

Rey, H.A. *The Stars: A new way to see them.* 1952. Reprint, Boston: Houghton Mifflin, 1962. *This is a really superb book for children ages 8–12. Shows the constellations as logical patterns, so that they can easily be seen and pointed out in the sky.*

Zim, Herbert S., and Baker, Robert H. *Stars: A golden guide.* Golden Press, 1951. Reprint, Racine, Wisconsin: Western Publishing Co., 1975. *After four decades, this little book still is hard to beat. Concise, profusely illustrated, and compact.*

## 24.2.9  Textbooks

Hartmann, William K. *Astronomy: The cosmic journey.* Belmont, Ca.: Wadsworth Publishing Co., 1978. *A thorough introductory college text.*

Pasachoff, Jay M. *Astronomy: From the Earth to the Universe,* 4th edn. Philadelphia: Saunders, 1989. *Thorough, well-written and illustrated college text.*

Seeds, Michael A. *Horizons: Exploring the Universe.* Belmont, Ca.: Wadsworth Publishing Co., 1981. *Unique text, written in a friendly and informal style.*

## 24.2.10 Magazines

*Astronomy.* 21027 Crossroads Circle, PO Box 1612, Waukesha, WI, 53187. A monthly magazine with excellent articles on all aspects of astronomy. Lively and well-written; perfect for adult beginners.

*Deep Sky*. 21027 Crossroads Circle, PO Box 1612, Waukesha, WI, 53187. Quarterly magazine with emphasis on observing objects outside the solar system. Includes a column by David Levy on variable stars.

*Odyssey*. 21027 Crossroads Circle, PO Box 1612, Waukesha, WI, 53187. Monthly magazine meant for children. Introduces the sky gracefully to the next generation.

*Sky and Telescope*. 49 Bay State Road, Cambridge, Mass., 02138. A superb monthly magazine with regular features devoted to astronomical news, observing, and telescope making. Includes a column by the author called 'Star Trails' about the amateur observing experience.

*Star Date*. RLM 15.308, The University of Texas at Austin, Austin, TX 78712. Beginner-aimed report and observing guide.

*The International Comet Quarterly*. Ed. Daniel W. Green, Smithsonian Astrophysical observatory, 60 Garden Street, Cambridge, Mass., 02138, USA. Since January, 1979, the *ICQ* has reported on comets and has maintained an archive of visual observations of comets.

# Index

A ring   108, 110
AAVSO, *see* American Association of Variable
   Star Observers
absolute magnitude   34
absorption spectrum   248
Academy of Sciences, Paris   131
Acapulco, Mexico   194
Achernar   7
Adam   270
Adams, John Couch   131, 132, 133,
   141
Adirondack Science Camp   29
Aetheria Darkening   121
afocal projection   229
Agfachrome   235
Airy, George   132, 133
Albategnius   71
albedo features, Martian   120
Albireo   5, 53, 165, 166, 170, 266
Albor   121
Alcock, George   41, 160
Alcor   165, 196
Aldebaran   7, 196
Algol   38, 168
Alkaid   42
Alpha Andromedae   6, 7
Alpha Capricornids   44
Alpha Centauri   7, 202
Alpha Coronae Borealis   196
Alpha Crucis   3
Alpha Draconis   232
Alpha Ophiuchi   196
Alpha Orionis, *see* Betelgeuse
Alpha Pegasi   6
Alphecca   5
Alpheratz   6
Alphonsus   71, 73, 76, 83, 86
Alps   71

Altair   5
   apparent magnitude   34
altazimuth mount   58, 232
aluminium   227
Amalthea   98
amateurs   xv
Ambartsumian, Viktor   198
American Association of Variable Star
   Observers   173, 177, 178, 182, 184, 185,
     186, 187, 188, 189, 190
'Anatomie of the World'   269
Andromeda Galaxy   8, 39, 206, 237
   *see also* Messier 31
Andromedids   46, 61
annual Andromedids   45
annular eclipses   240, 248
Antares   33
Antoniadi, Eugenios   99
Apennines   71, 74, 75, 76
aphelic chill   122
Apollo 11   74
Apollo–Amor group   142
Apollo project   65, 84, 86
apparent magnitude   34
apparition, Mars   117
Aquila   5, 8
AR Cephei   180
Arago, Francois   132, 133
Archimedes   71, 74, 75, 76
Arcturus   5, 165, 197, 206
   colour   35
Argelander, Father   180
Ariel   135
Aristarchus   73, 75, 83
Aristarchus of Samos   73
Aristillus   71, 74
Aristoteles   70, 83
Arizona   121

artificial satellites
  observing   32
  photographing   228
ashen light   128
Association of Lunar and Planetary
    Observers   66, 83, 84, 85, 86, 94, 95, 96,
    104, 109, 119
asteroids   139–47
  groups of   141, 142
  list of   143
  observing   142
  occultations   145
  photometry of   146
  rotation   146
Astronomical Almanac   85
Astronomical Ephemeris   112, 125
Astronomical Society of the Pacific   157
astronomy clubs   52–3, 262, 273
*Astronomy* magazine   37, 53, 142, 151, 247,
    273
astrophotography
  difficulties   226
  focusing   233
  guided   229–33
  unguided   226–7
Atlai Scarp   70
Atlas Stellarum   143, 172
atmospheric changes, Mars   119
Auriga   6, 7, 130, 271
Aurigids   43
aurora   29
Auroral Data Center   29
auroral report form   30
Autolycus   71, 74
averted vision   32, 39

B ring   108, 110, 111, 257
Baade   141
Bach   141
Bailly   86
Baily, Francis   245
Baily's beads   245, 247
Barlow lens   229
Barnard, E.E.   98, 156, 203
barred spirals   206
baseball   53
Battle Hymn of the Republic   27
Bayer, Johann   3
Beehive   198
  see also Messier 44
Bell Labs   60
belts   100, 104
Berlin Observatory   132, 139
Beta Andromedae   214
Beta Aurigae   196, 272
Beta Centauri   7, 202
Beta Ceti   87
Beta Crucis   7
Beta Cygni, see Albireo
Beta Lyrae   175, 182
Beta Orionis, see Rigel

Beta Tauri   271
Betelgeuse   7, 45, 179, 188
  colour   35
Beyer   154
*Beyond the Observatory*   34
Big Bang   61
Big Bear Lake, California   60
Big Dipper   1, 2, 4, 34, 43, 165, 196, 227, 266
Billy   75
binary system   256
binoculars   53–5
birds, migratory   66
bird-watching   53
'black drop' effect   129
Black Eye galaxy, see Messier 64
Blinking Planetary Nebula   203–4
BM Orionis   180
Bobrovnikoff Method   154
Bode, Johann   139
Bode's law   130, 140
Boeing 707   40
Bok, Bart   219
Bok globules   219
Bond, George   166
Bootes   5
bracketing exposures   228
Brahe, Tycho   72
Brahms   141
Brasch, Klaus   109
Brashear, Emily   272
Brashear, John   56, 272
Brashear camera   133
Breezy Hill   60
bridges   100
British Astronomical Association   161
Brooks, William R.   157
Burckhardt   70, 74
Burnham, Sherbourne Wesley   167

C ring   108, 110
Calgary, Alberta   3, 29
Calippus   74, 83
Callisto   98, 106
Cambridge   132, 141, 161, 174
*camera obscura* method   35
'canali'   114, 115
Canis Major   7
Canis Minor   6
Canopus   8
Capella   6, 7, 43, 87
  colour   35
  spectral type   35
Carina   8, 227
Carrington, Christopher   91
Cassini   72, 83
Cassini, Giovanni   72, 107
Cassini spacecraft   113
Cassini's Division   107, 257
Cassiopeia   6, 7
50 Cassiopeiae   232
Castor   7, 46, 166

cataclysmic variables   175, 179–80, 183
Catalina Mountains   51
Catharina   74
cats   164
Caucasus Mountains   70, 71, 74
Cavalerius   73
'Celestial Police'   140
Central Bureau for Astronomical
    Telegrams   149, 161, 162, 191
central eclipses   240
central meridian transits   105
Cepheids   175–6, 183
Cepheus   6, 7, 38, 211
Ceres   140, 142
Cetus   6, 7, 87
CH Cygni   180
Challis, James   132, 133
Chapman, Clark   146
Charon   136
Chaucer, Geoffrey   141, 269
Chi Cygni   176, 190
Chi 1 Orionis   87
children   261–8
China   141
Chinese sunspot observations   90
Christian VIII, King   157
chromosphere   246–7
Chryse basin   122
CHU   105, 253
Cicero   268
*Circulars*   151–2, 161
circumpolar   2
cirrus clouds   4, 28, 50, 61
city skies   1, 3, 207–10
Clark, Alvin   56
Clavius   72, 75, 83
Clavius, Christopher   72
Cleomedes   70, 74, 83
clouds   259
    Martian   122–3
    Venusian   128
cloudy night   148–9
Clownface Nebula   203–4
Coal Sack Nebula   8, 202, 203
coathanger   53
Coberg   155
Cobra's Head   73, 75, 83
coldest night   147
Collins, Peter   134, 191
Colombo   74
coma, *see* comets
Coma Berenicids   46
Coma-Virgo regions   159, 211, 214
Comet 1910a   158
Comet 1982i, *see* Comet Halley
Comet Arend–Roland   40
Comet Biela   46, 61
Comet Bradfield   62, 152
Comet Denning–Fujikawa   156
Comet Encke   45, 151
Comet Giacobini–Zinner   45

Comet Halley   41, 43, 45, 51, 70, 150, 162,
    195, 272
Comet Hartley–IRAS   150
comet hunting xviii   40, 155–62
Comet Ikeya–Seki   151, 157
Comet IRAS–Araki–Alcock   160, 216
Comet Kohoutek   150
Comet Levy 1987a   149
Comet Levy 1987y   62
Comet Levy 1988e   151
Comet Liller   62
Comet Mrkos   40, 158
Comet Okazaki–Levy–Rudenko   148, 235
Comet Pons–Winnecke   43, 157
Comet Shoemaker–Holt   151
Comet Sorrells   160
Comet Swift–Tuttle   42
Comet Tempel   271
Comet Tempel–Tuttle   46
Comet West   150, 163
Comet White–Ortiz–Bolelli   40
Comet Wilson–Hubbard   40
comets   42, 147–64
    awards for   156, 157
    brightness   4, 153
    coma   150, 154
    discovery   160
    during eclipses   247
    dust   150
    gas   150
    groups of   151
    hunting   155–62
    naming   162, 164
    observing   151–4
    tails   150, 155
Como, Perry   41
CompuServe   152
Cone Nebula   210
conjunction   37
constellations, *see individual constellations*
Copernicus   72, 74, 75, 76, 83
Copernicus, Nicolaus   72, 269
copying, film   234
'corona' (auroral form)   29
corona   239, 240, 245, 246–7, 248, 249
Corona Borealis   5
coronagraph   96
Cosmos   81
Crab Nebula, *see* Messier 1
craters
    during a lunar eclipse   251
    Martian   121
Crepe ring   108, 10
Crux   7
cusp caps   128
Cygnus   5, 204, 230
    nova in   180, 191
Cyrene   72
Cyrillus   70, 74

D ring   108

Daniel, Zaccheus 157
Danjon, A. 251
dark nebulae 202–3
darkening sky 244
d'Arrest, H. 131, 132, 196
Davies, Sir John 270
Davis, Don 146
Dawes, William Butler 170
De Divinatione 268
declination 9
deep sky 165–238
*Deep Sky Monthly* 222
deep sky objects 39, 195–238
Deimos 123
Delta Aquarids 41, 268
Delta Cephei 38, 175, 182
Delta Crucis 7
Delta Draconis 232
Delta Leonids 43
Delta Leonis 196
'Demon Star', *see* Algol
Deneb 5, 191
    absolute magnitude 34
Denebola 206
Denning, William F. 156
Denver 242
desert features, Martian 120
Deslandres 71, 72, 75
Dessau, Germany 90
diamond necklace 245, 248
diamond ring 245
Dickinson, Terence 235
Dione 112
Dionysius 74
'dirty snowballs' 150
distances in the sky 10
diurnal effect 71
Dobsonian 52
Donne, John 269
Donohoe Medal 157
Double Cluster 210
double stars 165–73
    observing xvii 168–73
    separation 169–70
    *see also individual stars*
downdrafts 100
Draco 10, 45
Draconids 45
20 Draconis 170
Drew, George 131
Dreyer, John Louis Emil 195–6
Drygalski 86
Duhalde, Oscar 40, 193
Duke of Milan 135
Dumbbell Nebula, *see* Messier 27
dust, *see* comets
dust storms, Martian 119–20
dwarf novae 179–80, 183
Dyer, Alan 214

E-ring 108

Earth, rotation 227
earthlight 69
earthshine 69
Echo 107
Eclipse Comet 247
eclipse party 250
eclipses, observing xvii
    solar 239–49
    lunar 249–53
eclipsing binaries 175, 183
Edberg, Stephen J. xvi, 31, 181, 192, 193,
    202, 209, 236, 240, 257
Edmonton 214
Einstein Observatory 175
Ektachrome 225, 228, 234, 237, 240, 246,
    257
Eliot, T.S. 162, 164, 181
elliptical galaxies 206
elongation 37
Elysium 121, 122
emission nebula 201
Enceladus 112
Encke, Johann 107, 132, 133
Encke's Division 107
environment, during solar eclipses 247
Epsilon Cephei 38
Epsilon Lyrae 169
'equal out' 154
equator 9
equatorial belts and zones
    Jupiter 101
    Saturn 108–9
equatorial mount 58
equinox 9–10
    Martian 119
Eratosthenes 72, 75, 83
Erfle eyepiece 58
Eridanus 7
Eskimo Nebula 203
Eta Aquarids 41, 43
Eta Carinae 194, 201, 202
Eta Geminorum 179
Eta Ursae Majoris 42
Eudoxus 70, 74
Eunomia 143
Europa 98, 254
Evans, Reverend Robert 192–3
evening star 37
extragalactic nebulae, *see* galaxies
eye protection 241
eyepiece projection 229
eyepieces 57–8


F ring 108, 257
Faber and Faber 162
Fabricus 74
faculae 93
Ferret, The 155
festoons 100
filar micrometer 172

film   27
  hypersensitizing   234
  processing   234
fireball   50
Flagstaff, Arizona   114
'flames'   29
Flamsteed, John   132
flares   93
flash spectrum   248
flying saucers   115
Fomalhaut   44
forming gas   234
Foucault, Leon   59
Fracastorius   74, 83
Frederick VI, King   157
Fujikawa   156
Furnerius   73, 74

g Herculis   183
G ring   108
Gagarin   141
galactic center   8
galaxies   39, 204–7
  brightness   4
  classification   206
Galilean satellites   106
Galileo   55, 59, 90, 98, 107, 269, 270
Galileo probe   100
Galle, Johann   132
Gamma Andromedae   169, 266
Gamma Bootis   176
Gamma Crucis   3
Gamma Leonis   169, 266
Gamma Lyrae   175
Gamma Pegasi   7
Gamma Ursae Majoris   5
Gamma Virginis   166, 169
Ganges 'Canal'   121
Ganymede   98, 106
garden telescope   59
Gardner, W.H.   271
Garneau, DeLisle   66
Garnet Star   191
gas, *see* comets
Gassendi   73, 75, 83
Gassendi, Pierre   75, 107
Gauss   73
gegenschein   32
Gemini   7, 130, 203, 271
Geminids   41, 46
Geminus   70, 74
*General Catalogue of Nebulae*   196
Geographos   142
globular clusters   198–200
  classification   200
Goodricke, John   38, 168
'Grand Chercheur'   156
grandfather, observing with   87
granulation   93
grazing occultations   256
Great Nebula in Orion, *see* Messier 42

Great Red Spot   101, 105–6
Great Rift   8, 203
Great Square of Pegasus   6
Greeks   141
Green, Daniel W.E.   153, 155
Greenberg, Rick   146
greenhouse effect   126
Greenwich   92–3
Gregorian calendar   72
Grierson, Sir Herbert   269
Grimaldi   73, 75
Grimaldi, Francesco Maria   73
*Gulliver's Travels*   123
Gum Nebula   204

H Geminorum   130
Haemus Mountains   70, 71, 74
Hahn   74
Hall, Asaph   123–4
Halley   71
Halley, Edmond   195, 197
haloes   28
*Hamlet*   269
Hansteen   75
Harding, Karl   140
Hardy, Thomas   249
Harvard   166
Havard College Observatory   102, 120, 127,
  129, 134, 136
Harvard designation   180–1
hat trick   233
Hausen   86
Hawaiian volcanoes   121
Hazelbrook, New South Wales   192
Hektor   141
Hell   72
Hellas   121, 122
Heraclides Promontory   72, 75
Heraclitus   75
Hercules   198
Hercules Globular cluster   137
Herodotus   73, 83
Herschel, Caroline   114, 156
Herschel, John   8, 196, 200, 202
Herschel, William   59, 99, 107, 114, 130–1,
  140, 156, 166, 167, 191, 195, 200, 271
Herschel Wedge   89
Hertzsprung–Russell graph   195
Hevelius   73, 75
Hevelius, Johannes   75
Hill, Rik   95
Hipparchus   71
Hodgson, Ralph   273
Hopkins, Gerard Manley   61, 216, 271, 272
Horrocks   71
Horsehead Nebula   196
Houghton, W.   271
Houston, Walter Scott   40, 158, 219, 268
Hoyt, W.   99
H2 regions   192
Hubble, Edwin   206

Hubble Space Telescope   60
hub-cap   60
Hughes, M.   270
Humboldt   69, 73
Hunter, Tim   33, 197, 199, 205, 211, 213, 215
Hussey, Rev. T. J.   132
Hutchinson, T.   271
Huygenian   58
Huygens   121
Huygens, Christiaan   106, 107
Huygens space probe   113
Hyades   7, 180, 196, 210
Hyblaeus Extension   121
Hydra   157
hydrogen-alpha filter   96
Hyperion   112
hypersensitized film   234

Iapetus   112, 113
IC 434   196
Icarus   142
Ikeya, Kaoru   157
*In Memoriam*   271
Independence day   41
*Index Catalogue*   196
inferior planets   33–4
Infrared Astronomical Satellite   162
Ingalls, Albert   59
'in–out'   153
insect repellant   4
instrument problems   259
International Astronomical Union   68, 149, 151–2
*International Comet Quarterly*   153, 155
International Halley Watch Near Nucleus Studies Network   272
International Occultation Timing Association   254, 256
International Standards Organization   225
International Ultraviolet Explorer   175
Io   98, 254
IOTA, *see* International Occultation Timing Association
Iota Aurigae   271, 272
Iota Ceti   87
irregular galaxies   206
island universes, *see* galaxies
ISO, *see* International Standards Organization
Italy   272

Janssen   70, 74
Jarnac Observatory   62
*Jealous Heart*   194
Jedicke, Peter   27, 195
Jones, Albert   194
June Draconids   43
Juno   140, 142
Jupiter   27, 33, 53, 56, 63, 98–106, 266, 268, 271

and asteroids   141
and comets   151
drawing   102–3
observing in daylight   37
occultations   254, 258
Jura Mountains   72, 75

'*k*' factor   91
Kappa Cygnids   44
Kappa Orionis   7
Karl Marx   141
Kellner eyepiece   58
Kepler   73, 74, 83
Kepler, Johannes   73, 90, 141
Kerns, John   222
Kirch, Gottfried   155
Kitt Peak   60, 69
Kitt Peak National Observatory   76, 146
Kochab   46
Kodachrome   234
Kodak   96, 118, 227
    technical pan 2415   237
Konica   228
Kreutz group   151
Krueger, R.   270
Kuiper Airborne Observatory   60
Kuiper, Gerard   115

'Lady Windermere's Fan'   269
Lagoon Nebula   196
    *see also* Messier 108
Lagrangian points   141
Lalande, Joseph   132, 156
Lambert   72, 83
Langrenus   69, 74
Lansberg   72
Laplace Promontory   72, 75
Laputians   123
Large Magellanic Cloud   192, 193, 205
Latin   106
latitude   2
Leavitt, Henrietta   175
Lemay, Damien   230
Leo   6, 45, 193, 206
Leonids   42, 45
Leverrier, Jean Joseph   132, 133, 141
Lexell   131
liberation   70–1, 86
Licetus   75
light year   34
lightning   28
Lilienthal Observatory   140
limb haze   123
Ling, Alister   62, 218
Linné   74, 83
Little Dipper   5
Little Dumbbell, *see* Messier 76
London, England   3, 29
longitude   3
Louis XV   155
Louisiana State University   66

Lowell Observatory   55, 197
Lowell, Percival   114, 121, 125, 127, 133–4,
    136, 263
Lubbock, Constance   114
Lumicon   234
Luna Incognita   86
lunar eclipses   249–53
    colour change during   251
    darkness of   251
    photography   252
Lunar Meteor Search   66
lunar occultations   253–9
Lunar Orbiter   86
*Lunar Photoelectric Photometry Handbook*   83
lunar transient phenomena   81–4
Lynn, Loretta   194
Lyra   5, 203
Lyrids   43

M objects, *see Messier numbers*
McAuliffe   141
McDonald Observatory   115
McGill University   244
Machholz, Don   159, 222
MacKenzie, N.H.   xix, 271
McMath Solar Telescope   60, 76
magnetic poles   29
magnitudes   4, 34
Manilius   83
Manly, Peter   60
Mare Acidalium   117
Mare Crisium   69, 70, 73, 74, 83, 250
Mare Fecunditatis   69, 70, 72, 73
Mare Frigoris   72
Mare Humboldtianium   69, 74
Mare Humorum   73
Mare Imbrium   71, 72, 74, 75
Mare Nectaris   70
Mare Nubium   75
Mare Serenitatis   70
Mare Spumans   72
Mare Tranquilitatis   70, 74
Mare Undarum   72
Mare Vaporum   71
maria   37
    Martian   120
Mariner II   126
Mariner 9   120
Mars   33, 53, 62, 114–25, 263, 271
    drawing   118–19
    moons of   123–5
    observing   xviii
    rotation   117
Marseilles Observatory   156
Masefield, John   27
Massachusetts   114
Mattei, Janet   174, 177, 178, 184, 185, 186,
    187, 189, 190
Mattei, Michael   116
Maurolycus   70
Meier, Rolf   159

Memorial Day weekend   60
Menelaus   70, 74
Mercury   33, 37, 129–30
    observing   130
    transits   130
Mercury Theatre   115
Merope Nebula   202
Mesalla   74
Messier, Charles   63, 131, 155, 193, 195, 200,
    212, 213, 269
Messier 1   204, 212, 214, 215
Messier 2   220, 225
Messier 3   208, 218
Messier 4   208, 219
Messier 5   218
Messier 6   208, 219
Messier 7   208, 219
Messier 8   196, 212, 219
Messier 9   218
Messier 10   218
Messier 11   208, 220
Messier 12   218
Messier 13   137, 198, 199, 208, 218, 237
Messier 14   219
Messier 15   210, 221
Messier 16   219
Messier 17   201, 212, 219
Messier 18   219
Messier 19   219
Messier 20   212, 219
Messier 21   219
Messier 22   208, 219
Messier 23   219
Messier 24   220
Messier 25   220
Messier 26   220
Messier 27   203, 220
Messier 28   220
Messier 29   220
Messier 30   221
Messier 31   8, 39, 210, 213, 214, 221
Messier 32   210, 221
Messier 33   39, 212, 221
Messier 34   221
Messier 35   179, 208, 216
Messier 36   208, 215
Messier 37   208, 215
Messier 38   208, 215
Messier 39   208, 215
Messier 40   216
Messier 41   216
Messier 42   39, 166, 198, 201, 203, 207, 208,
    215, 234, 237, 266
Messier 43   216
Messier 44   39, 208, 216
Messier 45   210, 215
Messier 46   216
Messier 47   216
Messier 48   216
Messier 49   217
Messier 50   216

Messier 51    137, 210, 218
Messier 52    221
Messier 53    217
Messier 54    220
Messier 55    220
Messier 56    220
Messier 57    203, 204, 208, 220
Messier 58    217
Messier 59    217
Messier 60    217
Messier 61    218
Messier 62    219
Messier 63    218
Messier 64    207, 217
Messier 65    217
Messier 66    217
Messier 67    216
Messier 68    218
Messier 69    220
Messier 70    220
Messier 71    220
Messier 72    220
Messier 73    221
Messier 74    221
Messier 75    220
Messier 76    221
Messier 77    221
Messier 78    210, 216
Messier 79    199, 200, 210, 216
Messier 80    219
Messier 81    208, 216
Messier 82    206, 208, 216
Messier 83    218
Messier 84    211, 218
Messier 85    217
Messier 86    211, 218
Messier 87    206, 210, 211, 218
Messier 88    217
Messier 89    218
Messier 90    218
Messier 91    217
Messier 92    218
Messier 93    216
Messier 94    208, 218
Messier 95    217
Messier 96    217
Messier 97    216
Messier 98    217
Messier 99    217
Messier 100    217
Messier 101    216
Messier 102    218
Messier 103    221
Messier 104    218
Messier 105    217
Messier 106    218
Messier 107    219
Messier 108    217
Messier 109    217
Messier 110    210, 210, 221
Messier hunting    212–21

Messier marathons    221–5
Messier objects    159
meteor showers    41
    observing    xvii, 46–50, 266–8
    photography    50, 228
meteorite    41
meteoroid    41
meteorology    28
meteors    40–1
Metis    143
Metius    74
Meudon    55
Mexico    125
*Midsummer Night's Dream, A*    135
Milky Way    8, 34, 53, 175, 191, 200, 202,
    203, 206
Miller, Robert    10
*Miller's Tale, The*    269
Millman, Peter    49, 50
Milton, John    270
Mimas    112
Minerva    62, 63
minor planets, *see* asteroids
*Minor Planet Circular*    143
Mira    176, 181, 182, 183, 186–7, 188
Mirach, *see* Beta Andromedae
Miranda    62, 135
Mizar    6, 34, 43, 165, 266
Monocerotids    42, 46
Montanari, Geminiano    38
Montreal    45, 49, 66, 88, 92, 96, 109, 143,
    158, 167, 191, 213, 249, 255, 257
Moon    9, 33, 65–87, 235, 236
    drawing features of    77–9
    east and west    68
    maps    67
    mountain heights    84–5
    observing    37, 56, 265
    phases    3, 37, 67
    photographing    79–81, 229
    photometry    86–7
    ray systems    86
    volcanism    84
    young crescent    69, 76
moondark    253
moonlight    69
moonquakes    84
moonshine    69
Morris, Charles    154
Morris Method    154
Mount Evans, Colorado    60
Mount Jennings Observatory    115
Mount Palomar telescope    237
Mount St Helens    251
Mount Wilson    60
Mountains of Mitchell    122
mounts, telescope    58
Mr Spock    141
Mu Cephei    191
Murphy's Law    257
mutual eclipses and occultations    106

mylar   89

Nagler eyepiece   58, 198
naked eye   27, 35, 37
'*Naming of Cats, The*'   181
National Oceanic and Atmospheric
    Administration   29
National Research Council (Canada)   29, 49,
    50, 268
natural pinholes.244
nebulae   39, 200–3
    *see also* galaxies; globular clusters; planetary
      nebulae
Neptune   271
    discovery   131–3
    observing   135–6
Nereid   136
*Neutron Star*   194–5
*New General Catalogue*   195–6
New Jersey   115
New Orleans   248
*New York Times, The*   113
New Zealand   194
Newton   86
Newton, Isaac   57
Newtonian reflector   57
NGC253   212
NGC869   210
NGC884   210
NGC2237   210
NGC2264   210
NGC2392   203
NGC2438   216
NGC4565   212, 223
NGC5128   209
NGC5694   199, 200
NGC6356   218
NGC6530   196
NGC6633   152
NGC6826   203, 212
NGC6946   211
Nix Olympica   121
North America Nebula   230
North Star, *see* Polaris
Northern Delta Aquarids   44
Northern Iota Aquarids   44
northern lights, *see* aurora
Northern Piscids   45
Northern Taurids   45
*Norton's Star Atlas* 67
notebooks   xvi
Nova Cygni   40
Nova Search Division   191
novae, searching for   40
Nu 1 and Nu 2 Sagittarii   166
Nye, Derald   146

Oberon   135
Observatoire de Paris   63, 132
*Observer's Handbook*   37, 38, 98, 142, 161, 214,
    254, 256

observing records   61–3
*Observing Variable Stars*   149
obsidian   60
occultations
    by the moon   253–6
    by and of planets   256–9
Oceanus Procellarum   72, 75
off-axis guider   231
*Old Possum's Book of Practical Cats*   162
Olympus Mons   121
O'Meara, Stephen James   102, 110, 120, 127,
    129, 134–5, 136
Omega Centauri   199, 208, 209
Omega Nebula, *see* Messier 17
Omicron Ceti, *see* Mira
Oort cloud comets   150
open clusters   196–8
Ophelia   269
opposition   33
    Mars   117
'Orchestra'   270
Orion   7, 196, 227
Orion Nebula, *see* Messier 42
Orionids   41, 45
Ottawa   50, 159
Otto Struve   76
'out–out'   154
Owl Nebula, *see* Messier   97

Padua   55
Pallas   140, 142
Palomar Observatory Sky Survey   149
*Paradise Lost*   270
parhelia   28
Paris   55, 60, 70
Parker, Don   105, 111, 124
parsec   34
partial eclipses
    of Sun   243–4
    of Moon   250
Patroclus   141
Pegasus   6
Peirce   70, 83
Peltier, Leslie C.   9, 157, 268
penumbra   96, 239, 249
penumbral lunar eclipses   253
Penzias, Arno   60
Perseids   41, 42, 44
    1966   49
    1978   267
Perseus   6, 168
Petavius   69, 74
Phaethon   46
Phobos   117, 123, 248
Phoebe   112, 113
Phoenix, Arizona   60
photography   225–38
    *see also* astrophotography
photosphere   243
Picard   70, 83

Piccolomini   70, 74
Pickering, E.C.   166
Pico   71, 83
pillar   28
Pioneer   108
Piton   71, 75, 83
Planet X   115
planetary nebulae   203–4
   classification   204
Planetary Science Institute   146
planets   9
   during eclipses   247
   motion of   32–3
   observing in daylight   37
   photography   229
   *see also individual planets*
*Planets X and Pluto*   99
Plato   71, 72, 76, 83, 86
Pleiades   7, 39, 45, 198, 202, 236
   *see also* Messier 45
Plinus   74
Plossl eyepiece   58
Pluto   115
   discovery   133–4
   observing   136–7
poets   268–73
polar axis   231–3
polar caps, Mars   117, 120, 122
polar regions
   Jupiter   101
   Mars   121–2
   Saturn   108–9
Polaris   5, 6, 232
policemen   98, 101, 259
Pollux   7
Pons, Jean Louis   131, 156
Pope   107
Pope, Alexander   135
Pope Urban VIII   55
Porter, Russell   59
Posidonius   83
Potsdam   168
Praesepe, *see* Messier 44
prime focus photography   225, 229, 231
'Princess, The'   40
Princeton University   157
processing film   234–5
Proclus   70, 83
Proctor, Mary   156
Procyon   6
prominences   96, 239, 246, 248
proper motion survey   197
Ptolemy   71
Ptolemy, Claudius   166, 197
Pulkovo Observatory   60
Purkinje effect   182
Pyrenees Mountains   74
Pythagoras   73, 76
Pytheas   72

Quadrantids   41, 42–3

quadruple stars, *see* double stars
quasars   207
Queen's University   272
Quintus Ennius   268

R Canum Venaticorum   177
R Ceti   186
R Coronae Borealis   180, 183, 188, 189
R Leonis   176, 188
R Leporis   188
radiant   42
rainbows   28
Ramsden eyepiece   58
*Rape of the Lock, The*   135
Raphael   270
'rayed arc'   29
Reber, Grote   60
reciprocity failure   225
red planet   263
reflection nebulae   202
reflector   52, 57
refractor   52, 55
Reid, William   157
retrograde   33
Rhaeticus   75
Rhea   112
Riccioli   75, 83
Riccioli, Johannes Baptista   75, 107, 128, 166
Rigel   7
   colour   35
right ascension   9
Rima Tenuis   122
Ring Nebula, *see* Messier 57
Riphaeus Range   72, 75
Riverside Telescope Makers Conference   60
RKE eyepiece   58
Robinson, F.N.   269
Roman Catholic Church   98
*Romance of Comets, The*   156
Rosenbaum, Gary   121
Rosette Nebula, *see* NGC2237
Royal Academy of Great Britain   123
Royal Astronomical Society   90, 130, 195
Royal Astronomical Society of Canada   37, 67,
   92, 96, 109, 142, 143, 161, 255
Royal Society (Great Britain)   57, 130
RR Coronae Borealis   183
RS Ophiuchi   190
RS Ursae Majoris   188
RU Herculis   189
RU Pegasi   191
Rudenko, Michael   134

S Persei   188
S Ursae Majoris   188
Sabine   74
Sagan, Carl   81
28 Sagittarii   108, 257
Sagittarius   5, 8, 166
Santa Claus   166
saros cycle   243

Saturn 9, 33, 106–14, 266, 271,
    drawings 109–10
    occultations 254, 257
    rings 100, 106, 108
    shape 107
Saturn IV-B 84
Scheiner 72
Schiaparelli 114, 121, 263
Schickard 73, 84
Schmidt cameras 230
Schmidt–Cassegrain reflector 57
schools 261
Schroter's Valley 73
Schur, Chris 148
Schwabe, Heinrich 90
*Scientific American* 59
scintillation 99
Scorpius 5, 8, 227
Scotti, Jim 62
'Sea Fever' 27
seasonal changes, Mars 119
secular changes, Mars 119
seeing 99–100, 168
seismographs on the Moon 84
Seki, Tstomu 157
*Selenographia* 76
Seleuchus 73
setting circles 10
Seven Sisters 39
shadow bands 244
Shakespeare 135, 141, 257, 269
Shapley, Harlow 34, 175, 198, 200
Shelton, Ian 193
'Shepheardes Calender' 268
Shklovskii, Iosef 117
Short 86
Shurz, Senator Carl 268
sidereal day 3
Sidgwick Method 153
Sigma Leonids 43
Sigma Octantis 3
Sigma 2816 169
Sigma 2819 169
Silverman, Karen 162
Sinus Iridium 72, 75, 83
Sinus Medii 71, 75
Sirius 7, 196, 197
*Skalnate Pleso* 161
Skiff, Brian 69
*Sky and Telescope* magazine 37, 38, 52, 68, 98,
    142, 151, 152, 170, 243, 247, 254,
    273
*Sky Atlas 2000.0* 161
Skyline 152
skytel 152
*Skyward* 109
Slipher, V. M. 133–4
Small Magellanic Cloud 175, 205, 206
Smith, Brad 170
Smithsonian Astrophysical Observatory 155
Society for Young Amateur Observers 156

Socratic method 263
solar day 3
solar eclipses 239–49
Solar Maximum Mission 151, 247
Solar System Ephemeris 85, 119
Solis Lacus 117, 119, 120, 122, 125
Solwind 151, 247
Sombrero Galaxy, *see* Messier 104
'Song of Honour' 273
songs 27
Sorrells, William 160
South African railroad workers 158
south tropical disturbance 101
Southern Cross 7, 8, 202
Southern Delta Aquarids 44
Southern Iota Aquarids 44
southern lights, *see* aurora
Southern Piscids 45
Southern Taurids 45
Soviet Union 32
space shuttle 32
spectra during eclipses 248
spectral type 35
spectroscope 35
Spenser, Edmund 268
Spica 5
    colour 35
spiral galaxy 206
Spitzbergen Mountains 75
Springfield mount 59
Springfield, Vermont 60
SS Cygni 175, 179–80, 181
Stange, G. 271
star charts 10–26
'Star Gazers' 271
star party 1
star trails, photographing 227–8
*Starlight Nights* 157
Star–Queen Nebula, *see* Messier 16
*Starry, Starry Skies* 27
stars
    brightness 34
    colour 35
    during eclipses 247
    *see also* double stars; *individual star names*;
        variable stars
Stellafane 59–60
Stevinus 73
Straight Range 72
Straight Wall 71
Struve, Otto 167, 171
Struve, Wilhelm 167
summer camp 1
summer triangle 5
Sun 3, 4, 87–97
    absolute magnitude 34
    colour 35
    drawing 94–6
    and filters 88
    observing 35, 60, 89, 264
    observing form 92

photographing   96–7, 228–9
  spectral type   35
sun filter   35
sundial   59
sundog   28
Sunflower, *see* Messier   63
Sungrazers   151
sunspot maximum   240, 247
sunspot minimum   246
sunspot number   91
sunspots   35, 56, 66, 88
  groups of   93
  observing   90–3
  observing form   36
superior planets   33
supernova of 1572   72
Supernova 1987A   40, 192–4
Surveyor   86
Swan Nebula   201
  *see also* Messier   17
Swift, Jonathan   123, 147
Syrtis Major   117, 119, 120–1, 123

T Cephei   180
T Coronae Borealis   189, 190
T Tauri   180
T Ursae Majoris   188
tails, *see* comets
tall ship   28
Tarantula Nebula   192
Taruntius   70, 74, 84
Tau Herculids   42, 43
Taurids   42, 228
teaching   262–8
telescope   55–7
  choosing   51–3, 207
  making   58
  mounts   58
temperate belts
  Jupiter   101
  Saturn   108–9
*Tempest, The*   135
Tenerife Mountains   72
Tennyson, Alfred Lord   40, 271
terminator   67, 78
terminator shadows   128
Tethys   112
Tewfik   247
Texas   115
Tharsis   121, 122
Theophilus   70, 74, 76, 84
Theta Orionis   166, 203, 266
'Thoughts on Various Subjects'   147
Thuban   10
thunderstorm   253
time   3
time and temperature method   235
*Times, The*, of London   272
timing occultations   254
Timocharis   84

Titan   107, 112–13, 257
Titania   135
Tithonius Lacus   120
Titius, Johann Daniel   139
Titius–Bode Law   140
Tombaugh, Clyde   115, 133–4, 137, 141, 157,
  170, 197, 200
Tombaugh–Smith seeing scale   170
total eclipses   240, 244–5
transits   241
  of Mercury   130
  of the Sun by Phobos from Mars   248
  of Venus   128–9
'Treatise on the Astrolabe'   269
Triangulum   39
Triassic period   206
Trifid emission nebula, *see* Messier   20
triple stars, *see* double stars
Triton   136
Tri-X   228
Trojans   141
47 Tucanae   199, 208
Tucson   39, 51, 162, 197, 199
Tuthill, Roger   157, 232
Tycho   72, 73, 74, 76, 84, 195

umbra   96, 249
Umbriel   135
United States   32
United States Naval Observatory   123
updrafts   100
*Uranometria*   3
*Uranometria 2000.0*   161, 172
Uranus   271
  discovery   130–1
  observing   134–5
  rotation   134–5
urban sites, *see* city skies
Ursa Major   6
Ursa Major Cluster   197
Ursa Major Group   196
Ursa Minor   5
Ursids   46

V Bootis   176–8, 188
Valis Marineris   120
variable stars   38, 173–95
  during eclipses   247
  naming   180–1
  observing   xvii, 182–91
V1500 Cygni   180
Vega   5, 10, 43, 165
  absolute magnitude   34
Veil Nebula   204
Vendelinus   69, 74
Venus   4, 33, 53, 56, 125–9, 263, 266
  observing in daylight   37, 126–7, 264
  phases   126
  transits   128–9
Verichrome Pan   228
Vermont   41

Vienna Academy of Sciences   157
3C273 Virginis
Virgo   5, 193, 207
Vogel, H. C.   168
Voroncov–Velyaminov scheme   204
Voyager   98, 101, 102, 108, 112, 134, 135,
   136
Vulcan   90

W Coronae Borealis   184
W Herculis   184
W Orionis   185
Walter   71, 72
*War of the Worlds*, 115, 263
Ward, Dan   28, 152, 201, 205, 211
Wargentin   73, 75
Warner, H. H.   156, 157
weather   4
Weidenschilling, Stu   146
Welles, Orson   115, 263
Westfall, John   83
'When I heard the learn'd Astronomer'   273
Whipple, Fred   150
Whirlpool Nebula   137, 205
   *see also* Messier   51
Whitman, Walt   273
Wild Duck, *see* Messier   11
Wilde, Oscar   269
*Wilhelm Struve Catalog*   171
Williamson, Isabel   49

Wilson, Robert   60–1
Wilson, Stewart   40
wind   259
Winnipeg   241
Wittenburg   139
Wolf, R.   91
Wordsworth, William   271
Wratten filters   101, 118, 122, 123
WWV   105, 253, 254

X Cameleopardalis   188
X Geminorum   188
X Herculis   183–4
Xi Ursae Majoris   172
X-rays   175
   emissions from dwarf novae   179

Yanaka   160
Yerkes Observatory   55, 60

Zagut   84
Zeta Cephei   38
Zeta Hercules   172
Zeta Lyrae   175
Zeus   98
Zigel, Feliz   124
zodiacal band   32
zodiacal light   31
zones   100
Zurich, Switzerland   91